Lecture Notes
in Control and Information Sciences 405

Margareta Stefanovic and Michael G. Safonov

# Safe Adaptive Control

## Data-Driven Stability Analysis and Robust Synthesis

**Authors**

Prof. Margareta Stefanovic

University of Wyoming
Dept. Electrical & Computer Engineering
1000 E. University Ave.
Laramie, Wyoming 82071
USA
E-mail: mstefano@uwyo.edu

Prof. Michael G. Safonov

University of Southern California
Dept. Electrical Engineering
3740 McClintock Ave.
Los Angeles, California 90089
Hughes Aircraft Electrical Engin.Bldg.
USA
E-mail: msafonov@usc.edu

ISBN 978-1-84996-452-4 e-ISBN 978-1-84996-453-1

DOI 10.1007/978-1-84996-453-1

Lecture Notes in Control and Information Sciences ISSN 0170-8643



*Typeset & Cover Design:* Scientific Publishing Services Pvt. Ltd., Chennai, India.

Printed in acid-free paper

5 4 3 2 1 0

springer.com

*To the memory of my father*

Margareta Stefanovic

*Dedicated to my loving wife, Janet*

Michael Safonov

# Preface

This monograph is intended to present a unifying theory of adaptive control, providing a theoretical framework within which all adaptive methods and iterative learning algorithms can be interpreted, understood and analyzed. The concept of the unfalsified cost level central to this theory leads to a simple framework for understanding the causes and cures for model mismatch instability. The need for an encompassing theory such as this stems from the control and stability issues of highly uncertain systems. After decades of adaptive control development, the opinions in the research community are still divided over what assumptions need and must be made about the controlled plant and its environment in order to assure stable operation, and hence resulting robustness of adaptive control algorithms. The study presented here unifies the results obtained in the last decade in developing a working theory of the *safe adaptive control*, that is, control with stability guarantees under a minimal knowledge of the plant and operating conditions, and as few prior assumptions as possible.

We hope that the book will prove to be a helpful means to learn and understand the parsimonious approach to the problem of the uncertain systems control. It can be used as a reference book on adaptive control with applications for readers with a background in feedback control systems and system theory. Given that uncertainty is one of the main features of many complex dynamical systems, the main objective of the book is to provide a concise and complete treatment of the robustly adaptive control for such highly uncertain systems. A.S. Morse's work [67] on hysteresis switching in parameter adaptive control played an important motivational role for this work. Since the publication of that study, a number of results have been reported in the "multiple model" switching adaptive control (also known in the literature as *identification based methods*), and "multiple controller" switching adaptive control (also known as *performance based methods*). The interest has been stimulated by the fact that all real-world systems are prone to unexpected modes of behavior, stemming either from the variations in their structure, parameters, disturbances, or unanticipated changes in their working environment. Control of the plant under such precarious conditions is a notable challenge, as the majority of the control methods are based on an identified model of the plant, with stability and performance conditions being closely tied to the modeling assumptions. However, a fundamental difficulty related to the model based approach is the fact that the modeling uncertainties are always present, stemming from intrinsic nonlinearities, unmodeled disturbances, simplifications, idealizations, linearizations, model parameter inaccuracies *etc.*

The results of the book are divided into two major entities. In the first part, we provide theoretical foundation for the robustness stability and convergence analysis in the general switching control setting, by allowing the class of candidate controllers to be infinite so as to allow consideration of continuously parameterized adaptive controllers, in addition to finite sets of controllers. We show that, under some mild additional assumptions on the cost function (chosen at the discretion of the designer, not dependent on the plant), stability of the closed loop switched system is assured. Additionally, in the cases when the unstable plant modes are sufficiently excited, convergence to a robustly stabilizing controller in finitely many steps is also assured. We emphasize the importance of the property of the cost functional and the candidate controller set called "cost detectability". Also provided is the treatment of the time-varying uncertain plants, and the specialization of the general theory to the linear time-invariant plant, which provides bounds on the size of the closed loop state.

In the second part, we provide simple but incisive examples that demonstrate the pitfalls of disregarding cost detectability in the design of the control system. The examples demonstrate that, when there is a mismatch between the true plant and the assumptions on the plant ("priors"), a wrong ordering of the controllers can in some cases give preference to destabilizing controllers. This phenomenon is known as *model mismatch instability*. At the same time, we emphasize that prior knowledge and plant models, when they can be found, can be incorporated into the design of the candidate controllers, to be used together with the safe switching algorithm. Other examples refer to the applicability of the safe adaptive control theory in improving performance of the stable closed loop system.

The book is written for the researchers in academia and research centers interested in the treatment of achieving stability and performance robustness, despite the wide uncertainties frequently present in real life systems. Graduate students in engineering and mathematical sciences, as well as professional engineers in the wide field of adaptive control, supervision and safety of technical processes should find the book useful. The first two groups may be especially interested in the fundamental issues and/or some inspirations regarding future research directions concerning adaptive control systems. The third group may be interested in practical implementations which we hope can be very helpful in industrial applications of the techniques described in this publication.

This book unifies the results of the Ph.D. thesis of the first author, and the seminal unfalsified control theory introduced into the adaptive control literature by the second author in the mid-1990s, as well as the examples derived by the authors and their coworkers.

We wish to thank Professors Edmond Jonckheere, Leonard Silverman, Petros Ioannou and Gary Rosen, who served on the thesis committee of the first author, for their detailed comments and suggestions. We also wish to thank Prof. João Hespanha for stimulating discussions and referral to the hysteresis switching lemma of Morse, Mayne and Goodwin and the related work.

The first author wants to express gratitude to Prof. Michael Safonov, for providing excellent guidance, encouragement and support during the course of her doctoral research. She is also indebted to Prof. Edmond Jonckheere with whom she has had the opportunity to collaborate on a number of relevant research issues. Thanks are also due to her former doctoral student Jinhua Cao, for his contribution to some of the concepts and examples presented in this work. Thought-stimulating conversations with Professors Suresh Muknahallipatna (University of Wyoming) and Brian D.O. Anderson (Australian National Unversity) have influenced the interpretation given to some of the results.

The second author expresses his sincere appreciation to the many colleagues and students whose contributions have helped shape the development of the unfalsified control concept that is the heart of our theory of adaptive control. Former students who have made direct contributions to the theory include Ayanendu Paul, Crisy Wang, Fabricio Cabral, Tom Brozenec, Myungsoo Jun, Paul Brugarolas, and especially, Tom Tsao who helped lay the foundations of the unfalsified control theory. We also thank Robert Kosut, Brian D.O. Anderson, Lennart Ljung, Michel Gevers, Mike Athans, Charles Rohrs, Masami Saeki, Petar Kokotovic, Petros Ioannou, Jeroen van Helvoort, Karl Åström, Bob Narendra, Jan Willems and Maarten Steinbuch whose work and words helped to shape our perspective and focus our thoughts on the need for a theory of safe adaptive control.

We are thankful to the Air Force Office for Scientific Research for partially supporting our work leading to the completion of this book (grant F49620-01-1-0302). The monograph is the result of the work conducted at the University of Southern California and the University of Wyoming.

Finally, we would like to thank the team of Springer publications, Mr. Oliver Jackson and his editorial assistants Ms. Aislinn Bunning and Ms. Charlotte Cross of Springer London, for their support and professionalism in bringing out the work in the form of monograph in an efficient manner.

Laramie, WY,
Los Angeles, CA
July 2010

Margareta Stefanovic
Michael G. Safonov

# Contents

# Part I
# Overview of Adaptive Control

# Chapter 1
# Overview of Adaptive Control

**Abstract.** In this chapter, we discuss the problem of the adaptive control and its present state of the art. We concern ourselves with the limitations of the adaptive control algorithms and discuss the tradeoff between the restrictions and benefits introduced by prior assumptions. Main contributions to the adaptive control within the last several decades are briefly reviewed to set up the stage for the current research results. We conclude the chapter by discussing the contribution of the monograph, which provides a solution to the long standing problem of model mismatch instability, cast in a parsimonious theoretical framework.

## 1.1 Discussion of the Adaptive Control Problem

A defining notion of an adaptive system in engineering is the process of adaptation, which not unexpectedly finds its place in many other diverse disciplines, such as biology, economics, and operations research. One of the earliest is the use of this notion in biology, in which adaptation is considered to be a characteristic of an organism that has been favored by natural selection and increases the fitness of its possessor [36]. This concept is central to life sciences, implying the accommodation of a living organism to its environment, either through physiological adaptation (involving the acclimatization of an organism to a sudden change in environment), or evolutionary adaptation, occurring slowly in time and spanning entire generations. The definition of this notion found in the Oxford Dictionary of Science, namely "any change in the structure or functioning of an organism that makes it better suited to its environment", lends itself promptly not only to living organisms but also to any entity endowed with some form of a learning ability and a means to change its behavior. In control theory, the term was introduced by Drenick and Shahbender in [28] (further elaborated in [101]), which refers to a system that monitors its own performance in face of the unforeseen and uncertain changes and adjusts its parameters towards better performance and is called an *adaptive control system*. As discussed in [76], many definitions of adaptive control systems have appeared during the course of the last fifty years, all pointing to some property of the system

M. Stefanovic and M.G. Safonov: Safe Adaptive Control, LNCIS 405, pp. 3–12.
springerlink.com 

considered to be vital by the proponents of the definition. An insightful discussion in [76] reveals that the difficulties in reaching a consensus on the definition of adaptive control systems can be summarized by the observation that "adaptation is only in the eyes of the beholder".

Traditional adaptive control is concerned with the plant parameter uncertainty, which impairs the available feedback. Parametric adaptation is defined as the case when the parameters of the system under consideration or the parameters of the noise signals are insufficiently known, whereas the structure of the system is well understood. In [76], such adaptive systems are contrasted with the so-called 'switching systems', defined as the systems where structural changes take place because of the modifications in the interconnection between subsystems. Based on such delineation between parametric and structural adaptive systems, the theory of traditional adaptive control proceeds along the lines of parametric adaptive systems, using continuous dynamical adjustment of the control parameters to converge to a system with preferred performance. This process is characterized by the so-called 'dual control' actions [31]. As argued in [76], understanding and controlling the world are two closely related but distinct activities. In the case of adaptive systems as defined by Bellman and Kalaba, neither action alone is sufficient to cope with an imprecisely known process. Thus, it requires simultaneous acquisition of the process knowledge (through process estimation and identification) and determination of the necessary control actions based on the acquired knowledge. The two activities hold a complex relationship, and many diverse classes of adaptive control systems have been proposed and analyzed in the past several decades assuring stability of adaptive control systems under the dual control actions. In classical adaptive control terms, adaptation is a procedure involving either an online identification of the plant parameters or a direct tuning of the control parameters, in order to change the compensator, prefilter and the feedback path accordingly.

Changes in the basic paradigm of adaptive control took place some time in the late 1980s and beginning of 1990s, through more thorough consideration of the systems exhibiting vast variations in their structure or parameters, thereby rendering the traditional continuously tuned adaptive control methods too slow and ineffective. A need was recognized for controlling such systems in a discontinuous way, *i.e.*, by switching among a set of different controllers, each of which is supposed to satisfactorily control a different plant. The rationale for such a control action is based on the assumption that at least one of these controllers is able to control the given plant. Even though this assumption is not explicitly stated in the traditional adaptive control paradigm, it is nevertheless present there, too [76]. Such discontinuously controlled systems can be classified as 'switching', as described in the traditional adaptive control terms from the previous paragraph, but their switching nature does not necessarily stem from the changes in the interconnection between the plant subsystems. In fact, the plant itself may have a fixed but unknown structure and/or parameters, as in the case discussed above, but it becomes a switched system when connected with the members of the controller pool. A new light was cast on the the meaning of the *dual control* by the *data-driven unfalsified control theory*, which forms the basis for the results in this book. In this viewpoint, the part of the dual control dealing with the acquisition of the process knowledge should be seen as learning about the process through the behavior of its signals, rather than as identification of the mathematical model of the process and estimation of its parameters. Of course, mathematical models of the plant do have an important role in designing the controllers constituting the candidate pool from which to choose a controller that will be closed in feedback with the plant. This choice (taking the control action) depends on the acquired plant knowledge, *i.e.*, the knowledge of the plant behavior evidenced by its relevant signals. The implication of these ideas is the formulation of a much less restrictive framework for understanding and control of uncertain systems (with less dependence on the validity of the identified model). With this explanation in mind, we state that the *safe adaptive control* based on the unfalsified control notions is not merely another competing method in adaptive control defined by its different assumptions. The concept of unfalsified cost levels presented in this book yields a unifying theory of adaptive control. It provides a theoretical framework within which all adaptive methods and iterative learning algorithms can be interpreted, understood and analyzed. Most significantly, the unfalsified cost level concept provides a simple framework for understanding the causes and cures for the model mismatch instability. The apparent disparity between the continuously tuned traditional adaptive systems and switching adaptive systems vanishes if one considers that the continuously tuned control parameters of the traditional theory can be seen as a special case of the controller switching (with infinitesimally small dwell time between the subsequent switches). This general view of an adaptive control system, in which the control action is based on the learned characteristics of the process (plant) is depicted in Figure 1.1.

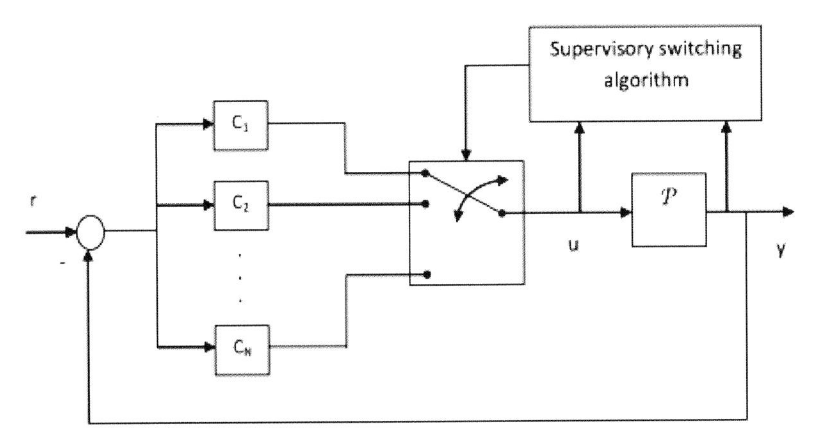


**Fig. 1.1** Adaptive control architecture consisting of an adjustable/switching controller block and a supervisory controller block

### 1.1.1 Brief History Overview

Control theory has traditionally been based on the intuitively appealing idea that a controller for the true process is designed on the basis of a fixed plant model. This technicality implies that one needs to acquire a reasonably good model of the plant, which would remain to reside within an acceptable range of accuracy for the duration of the plant operation. That an accurate, fixed model may be impossible to obtain, for example, because of the changing plant parameters or dynamics, or simply be difficult to verify (because of the complexity of the plant) was recognized sometime during the 1950s, notably during the aeronautical efforts to increase aircraft speed to higher Mach numbers in the presence of highly dynamic operating conditions such as varying wind gusts, sudden maneuver changes *etc.*, during the design of autopilots and stability assist systems for high-performance aircraft, and later on during the space exploration era. Adaptive control research thus emerged as an important discipline. As it often happens in much of the scientific research, the research advances in this area have followed a convoluted and sometimes erroneous path, and in certain cases led to severe consequences. In 1967, adaptive flight control system contributed to the crash of the hypersonic research aircraft X-15, causing the aircraft to oscillate and finally break apart. Though it was an isolated incident, the failure of the adaptive flight control system cast a doubt on its practical applicability. In the many years that followed, significant advances in nonlinear control theory were achieved, eventually bringing about the success in the onboard adaptive flight control of the unmanned unstable tailless X-36 aircraft, tested in flight in the 1990s. The benefits of the adaptation in the presence of component failures and aerodynamic uncertainties were clearly recognized. However, owing to the many prior plant assumptions made in theory, the most important question related to stability and robustness assessment is still being resolved on an *ad hoc* basis, by performing the tests of the closed loop system for many different variations of uncertainties in the Monte Carlo simulations. It is known that the cost of these heuristic tests increases with the growing complexity of the system. The limitations inherent in the conventional adaptive controllers have been recognized, and the need for safe adaptive control algorithms in the face of uncertainties has been set as the current goal.

According to the most commonly adopted view, an adaptive controller can be characterized as a parameterized controller whose parameters are adjusted by a certain adaptive strategy (Figure 1.1). The way its parameters are adjusted gives rise to a variety of different adaptive control strategies. The most popular ones will be briefly reviewed in the next section.

## 1.2 Literature Review

Adaptive control strategies can be categorized according to whether the controller parameters are tuned continuously in time, or switched discontinuously between discrete values at specified instants of time (called switching instants). In the second case, switching can even be performed among controllers of different structures, if appropriately designed. Also, control strategies can be distinguished based on whether the stability results of the closed loop control system rely on an identified plant model, or are independent of the plant identification accuracy and other prior assumptions.

In the first category, the methods based on the continuously tuned control parameters encompass most of the adaptive schemes starting from the early MIT rule and sensitivity methods in the 1960s and including many of today's algorithms. Because there is a plethora of results, we will not review them in detail here, but briefly mention the timeline in which they appeared: the sensitivity methods and the gradient based MIT rule in the 1960s [80, 111]; Lyapunov theory based and passivity based design in the 1970s [62, 76]; global stability results of 1970s and 1980s [2, 6, 29, 38, 60, 71, 78]; robustness issues and instability [85], and robust modification of the early 1980s such as dynamic normalization, leakage, dead zone and parameters projection [51], [53], resulting in the robust adaptive control theory [52, 76, 83] *etc.*; nonlinear adaptive control developments of the 1990s, as well as various alternative techniques such as extended matching conditions leading to the adaptive backstepping control [59]; neuro-adaptive control; fuzzy-adaptive control and several others. All of these modifications to the existing adaptive control theory resulted in stability and performance guarantees in the situations when the modeling errors are small, or when the unknown parameters enter the parametric model in a linear way. The limitations of traditional adaptive control theory are discussed in an insightful manner in [3].

In the late 1950s, the presence of a single unknown parameter (dynamic pressure) in the aircraft dynamics applications spurred research efforts resulting in the gradient based adaptation method of Whitaker *et al.* [111], which became known under the name "MIT rule". It is a scalar control parameter adjustment law proposed for adaptive control of a linear stable plant with an unknown gain. An approximate gradient-descent method is used in order to find the minimum of an integral-squared performance criterion.

Although practical and straightforward at first sight, the usability of the MIT rule turned out to be questionable; the performance of the closed loop control system proved to be unpredictable. Some explanations of the curious behavior were reported only decades after the inception of the concept.

The schematic of the plant controlled by the MIT rule is shown in Figure 1.2. The plant is represented by $k_p Z_p(s)$, where $k_p$ is the unknown gain with a known sign, and $Z_p(s)$ is a known stable transfer function representing the plant. The output of an adjustable positive gain $k_c(t)$ driven by a known reference input $r$ is chosen as the control input, and the value of the gain $k_c(t)$ is adjusted online based on the comparison between the plant output and the output of $k_m Z_p(s)$ driven by the same driving signal $r$. The gain $k_m$ is a known, predefined constant, with the same sign as $k_p$. The rationale behind this idea is the notion that, if $k_p k_c(t) = k_m \; \forall t$, it will result in a zero error $e(t) = y_p(t) - y_m(t)$. If, on the other hand, the error $e(t)$ is non zero, then an effort to continuously adjust $k_c$ to drive it to zero is undertaken.

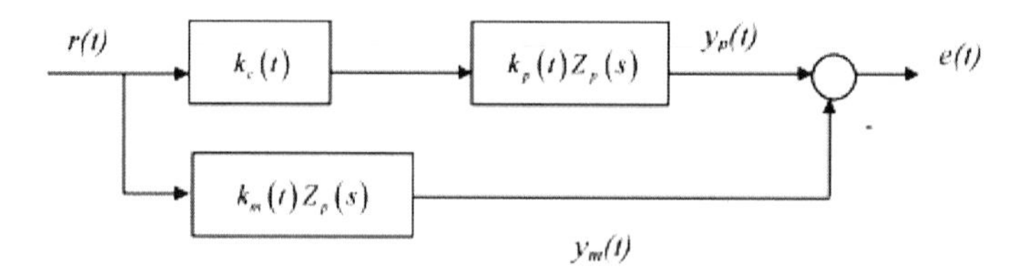


**Fig. 1.2** MIT rule

The MIT rule proposes the gradient descent (steepest descent) method for adjusting $k_c$:

$$\dot{k}_c = -\gamma \frac{\partial}{\partial k_c} J \tag{1.1}$$

where $J = \frac{1}{2}e^2(t)$ is the integral square error, and $\gamma$ is a positive gain constant representing the rate of descent. Since $e(t) = y_p(t) - y_m(t)$, one can also write:

$$\dot{k}_c = -\gamma(y_p(t) - y_m(t))y_m \,. \tag{1.2}$$

The problems with the MIT rule, despite its apparent usefulness and appealing simplicity, is that stability cannot be guaranteed, as varying levels of the gain $\gamma$ can sometimes result in satisfactory performance, and sometimes lead to instability. The explanation that followed decades later showed that the problems can be traced to the overlap between the time scale of the underlying closed loop dynamics and the time scale of the adaptive loop process.

In the first half of the 1990s, the idea of adaptive switching among distinct controllers has been introduced [37, 67, 75] to mitigate the drawbacks of conventional continuous adaptive tuning which occurred when the plant was faced with sudden or large changes in the plant parameters. It was recognized that the closed loop system's performance can deteriorate significantly, and even go into instability, when the plant uncertainty is too large to be covered by a single model, or when the changes in the plant parameters or structure are too sudden or large to be handled by the conventional on-line tuning. The motivation for introducing adaptive switching into the control system design was the desire to eliminate, or at least relax, the prior assumptions needed for stability that blighted the applicability of the existing control schemes. This gave rise to the switching supervisory control methods, where an outer loop supervisor controller decides which controller from the available pool should be switched on in feedback with the plant. In this regard, these methods bear similarity to the earlier gain-scheduling methods. However, unlike gain-scheduling methods, which use a pre-computed, off-line adjustment mechanism of the controller gains without any feedback to compensate for incorrect schedules, the recent switched adaptive control schemes incorporate data based logic for choosing a controller. The switching takes place at distinct time instants, depending on whether certain logic conditions have been met. The first attempt towards truly data-driven, almost-plant-independent adaptive control algorithms were proposed in [67] (later corrected in [68]) and then in [37]. In the former, it was shown that the only information required for stabilization is the order of the stabilizing linear controller ('universal controller'), provided that a stabilizing controller exists in a candidate set. The stabilization is then achieved by switching through the controllers in the candidate set until the stabilizing one has been found, and the switching logic is based on real time output data monitoring. The prior assumptions on which stability results are based are rather weak; however, practical applicability is limited because of the large transients occurring before the stabilizing controller has been found. In [37], another adaptive control strategy was proposed by Fu and Barmish, providing stricter results of stability and performance, but including additional requirements on the knowledge of the set of possible plants, compactness and an *a priori* known upper bound on the plant order. In both of these algorithms, the control parameters are adjusted on-line, based on the measured data; however, the switching follows a pre-routed path, without an inherent data based logic, which limits their applicability. Nevertheless, these methods paved a path towards truly data-driven adaptive control methods, unburdened by unnecessary prior assumptions in their stability analysis. They have been named 'performance based switching control algorithms' in [46] among others. In the 1990s, a notable further development in the data-driven adaptive control research was reported in [90], where it was shown that *unfalsified* adaptive control can overcome the poor transient response associated with the earlier pre-routed schemes by doing direct validation of candidate controllers very fast by using experimental data only, without making any assumptions on the plant beyond feasibility, and thus coming closer to a practical solution to the safe adaptive control problem. A related concept of the virtual reference signal (similar to the fictitious reference signal) was introduced in [19].

Concurrent research in a different vein produced results in the switching supervisory control using *multiple plant models* (also called 'estimator based switching control algorithms'), where the adaptive controller contains an estimator that identifies the unknown plant parameters on-line, and the control parameters are tuned using the plant parameter estimates at each time instant. These methods rely on the premise that the candidate model set contains at least one model which is sufficiently close (measured by an appropriate metric) to the actual, unknown plant (see, for example, [10], [61], [73]).

The preceding two assumptions, namely the assumption of existence of at least one stabilizing controller in the candidate pool needed in the 'performance based' control methods, and the assumption of existence of at least one model in the candidate model pool, which sufficiently captures the actual plant model (needed in estimator based schemes), can appear to be similarly restrictive. The important distinction between them is in the way they are put to use. The former assumption (which we call 'feasibility of the adaptive control problem' in this book) is the most fundamental assumption that has to be made if one expects to control a given system under given conditions. If it does not hold, then little can be said about future operation of the closed loop system. If it does hold, then the algorithm based on this assumption places no restrictions on the controller selection other than what the measured plant data indicate. Hence, it is a fundamental and minimal assumption. On the other hand, the assumption of at least one model in the candidate set, which captures the actual plant model may be deceptive. The reason is that the controllers designed for a model that is not very close to the actual plant may very well be stabilizing for the actual plant; however, they may be overlooked and discarded by an algorithm whose logic condition is based on the plant-model matching.

The research thrust in model based switching adaptive control of the last twenty years has produced incisive and important results, with many of them focusing on specific topics falling in the broad area of adaptive control. In [74], detailed results are given on the set-point adaptive control of a single-input, single-output process admitting a linear model. In that study, it is stated that, while the non-adaptive version for the set-point control problem is very well understood, the adaptive version is still rudimentary because the performance theories are not developed for the adaptive version, and because the existing stability proofs are valid only under restrictive, unrealistic assumptions of zero noise and/or un-modeled dynamics. In addition, one of the factors impeding the progress of adaptive control is pointed out, which is the seemingly innocuous assumption that the parameters of the nominal model of the controlled plant belong to a continuum of possible values. This assumption stems from the origins of the parameter adaptive control in the nonlinear identification theory. For example, if $P$ is a process to be controlled, admitting a linear, single-input single output model, then the modern adaptive control typically assumes that $P$'s transfer function lies in a known set $\mathfrak{M}$ of the form:

$$\mathfrak{M} = \bigcup_{p \in \Pi} B(\nu_p, r_p)$$

where $B(\nu_p, r_p)$ is the open ball of radius $r$ centered at the nominal transfer function $\nu_p$ in a metric space, $\Pi$ is a compact continuum within a finite dimensional space and $p \mapsto r_p$ is at least bounded. Then, for each $p \in \Pi$, the controller transfer function $\kappa_p$ is chosen to endow the closed loop feedback system with stability and other preferred properties for each candidate plant model $\tau \in B(\nu_p, r_p)$. However, since $\Pi$ is a continuum and $\kappa_p$ is to be defined for each $p \in \Pi$, the actual construction of $\kappa_p$ is a challenging problem, especially if the construction is based on linear quadratic Gaussian (LQG) or $H^\infty$ techniques. Also, because of the continuum, the associated estimation of the index of the ball within which $P$ resides will be intractable unless restrictive conditions are satisfied, which are the convexity of $\Pi$ and the linear dependence of the candidate process models on $p$. The so-called "curse of the continuum" is discussed in detail in [74], and the ideas to resolve the problems induced by the drawbacks of the continuity assumption are proposed.

In this monograph, we adopt a different perspective of adaptive control and its most urgent problems. Rather than providing a specialized treatise on some specific problem in adaptive control, we postulate the problem of a *safe stable adaptive control* in a general framework, considering an arbitrary plant (nonlinear in general) under minimal prior knowledge. In addition, we tackle the continuity notion, but in a different light: when a continuum of controllers are considered as candidates for the controller set, how can one switch among them so as to ensure stability and performance specifications? This question has been a long standing problem in the adaptive control community.

Estimator based adaptive control, both as the traditional continuous adaptive tuning, and more recently proposed discontinuous switching, relies on the idea of the Certainty Equivalence, which is a heuristic idea advocating that the feedback controller applied to an imprecisely modeled process should, at each instant of time, be designed on the basis of a current estimate of the process, with the understanding that each such estimate is to be viewed as correct even though it may not be. As argued in [74], the consequence of using this idea in an adaptive context is to cause the interconnection of the controlled process, the multi-controller and the parameter estimator to be *detectable* through the error between the output of the process and its estimate for every frozen parameter estimate, and this holds regardless of whether the three subsystems involved are linear or not. Even though this idea is intuitively appealing and has been the basis for many known adaptive control designs, the reliance on the plant detectability notion as described above can be forgone and replaced with the less restrictive cost detectability notion.

The reference [44] proposes hysteresis based supervisory control algorithms for uncertain linear systems. An extension to the nonlinear case, albeit in the absence of noise, disturbances, and unmodeled dynamics has also been discussed. It is argued that the robustness issues in the nonlinear case may be handled using the nonlinear extension of the Vinnicombe metric. This reference also proposes a 'safe multiple model adaptive control algorithms', similarly as in the study presented here. The difference in these two definitions of what makes adaptive control safe is subtle but carries important implications. In [44] and the related references, a safe adaptive controller is stated as the one "capable of guaranteeing that one never switches to a controller whose feedback connection with the process is unstable". We adopt a substantially different paradigm in this study. Since the process to be controlled, as well as the operating conditions, are assumed to be unknown, one can never be certain *a priori* whether a particular controller will be destabilizing for the plant under arbitrary operating conditions. That is, one cannot be completely assured that a controller switched into the loop with the process is destabilizing, *unless* and until the data recorded from the process show that the controller is failing to stabilize the plant. Only then can one hope that the control scheme will be able to recognize the deteriorating behavior of the closed loop and discard the controller, in search for a different controller (an as-yet-unfalsified one). It is precisely that ability of the adaptive control, namely recognition of the closed loop behavior based on on-line monitoring of the output data, and the corresponding corrective action (switching to an alternative controller), which we seek to characterize and quantify, so that provably correct and fully usable adaptive control algorithms can be designed. Of course, ongoing research may steer attention to newly developed cost functions which may be more swift than the ones currently in use in recognizing the instability caused by the currently active controller.

## 1.3 Monograph Overview

This book discusses hysteresis switching adaptive control systems designed using certain types of $L_{2e}$-gain type cost functions, and shows that they are robustly stabilizing if and only if certain plant-independent conditions are satisfied by the candidate controllers. These properties ensure closed loop stability for the switched multi-controller adaptive control system whenever stabilization is feasible. The result is a safe adaptive control system that has the property that closing the adaptive loop can never induce instability, provided only that at least one of the candidate controllers is capable of stabilizing the plant. As reported in the early study of Vladimir Yakubovich [114], the concept of feasibility plays a key role in ensuring finite convergence, though the definition of feasibility in this book differs in that it requires no assumptions about the plant or its structure.

The contribution of this monograph is the presentation of a parsimonious theoretical framework for examining stability and convergence of a switching adaptive control system using an infinite class of candidate controllers (typically, a continuum of controllers is considered). This property of the candidate controller class is essential when the uncertainties in the plant and/or external disturbances are so large that no set of finitely many controllers is likely to suffice in achieving the control goal. We show that, under some non-restrictive assumptions on the cost function (designer based, not plant-dependent), stability of the closed loop switched system is assured, as well as the convergence to a stabilizing controller in finitely many steps. The framework is not restricted by the knowledge of noise and disturbance bounds, or any specifics of the unknown plant, other than the feasibility of the adaptive control problem.

This treatise presents data-driven control of highly uncertain dynamical systems under as little as possible prior knowledge, with the origins of uncertainty ranging from the time-varying parametric changes (both slow and abrupt), hard-to-model plant dynamics, deteriorating plant components/subsystems, system anomalies (such as failures in actuators and sensors), unknown/adverse operating conditions, *etc*. Applications are varied, such as safety critical systems area, aircraft stability systems, missile guidance systems, dynamically reconfigurable control for *ad hoc* battlefield communication networks, health monitoring/autonomous fault diagnosis/reconfigurable control in spacecraft systems, reconnaissance and rescue operations from the disaster areas *etc*. Those applications are affected by the time critical issues that prevent timely characterization of the plant and surrounding conditions, rendering most classical model based control strategies possibly inefficient. The methodology presented in this text is based on switching into a feedback loop one of the controllers in the available candidate set, where the switch logic is governed by the collected input/output data of the plant (either in open- or closed loop), and the proper choice of the performance index (or a cost function). Such methodology is named *Safe Adaptive Control*, and is essentially realized according to the paradigm of the Multiple Model/Controller Adaptive Control. Safety in the title refers to the attempt to control the plant and furnish stability proofs without regard to the assumed plant structure, noise and disturbance characteristics.

# Part II
# Safe Switching Adaptive Control

# Chapter 2
# Safe Switching Adaptive Control: Theory

**Abstract.** In this chapter, we lay out the foundations of the safe switching adaptive control theory, based on the controller falsification ideas. We study the properties of a closed loop switched adaptive system that employs an algorithm which prunes candidate controllers based on the information in the real-time measurement data. The goal of this pruning is to evaluate performance levels of the candidate controllers simultaneously before insertion into the closed loop. We formulate the list of the plant-independent properties of the cost function required for the stability guarantee. The crucial role of the cost detectability property is studied. We provide significant generalization by allowing the class of candidate controllers to have arbitrary cardinality, structure or dimensionality, and by strengthening the concept of tunability. Specialization to the linear time-invariant plants is discussed, giving insight into the bounds of the closed loop states. The extension to the time-varying plants is considered next, where cost detectability is shown to hold under modified conditions, and an upper bound on the number of switches over an arbitrary time period is derived.

## 2.1 Background

The book *Adaptive Control* [7] begins in the following way: "In everyday language, 'to adapt' means to change a behavior to conform to new circumstances. Intuitively, an adaptive controller is thus a controller that can modify its behavior in response to changes in the dynamics of the plant and the character of the disturbances".

Whether it is conventional, continuous adaptive tuning or more recent adaptive switching, adaptive control has an inherent property that it *orders controllers based on evidence found in data*. Any adaptive algorithm can thus be associated with a cost function, dependent on available data, which it minimizes, though this may not be explicitly present. The differences among adaptive algorithms arise in part due to the specific algorithms employed to approximately compute cost-minimizing controllers. Besides, major differences arise due to the extent to which additional assumptions are tied with this cost function. The cost function needs to be chosen

M. Stefanovic and M.G. Safonov: Safe Adaptive Control, LNCIS 405, pp. 15–40.
springerlink.com 

to reflect control goals. The perspective we adopt hinges on the notion of feasibility of adaptive control. An adaptive control problem is said to be feasible if the plant is stabilizable and at least one (*a priori* unknown) stabilizing controller exists in the candidate controller set, which achieves the specified control goal for the given plant. Given feasibility, our view of *a primary goal of adaptive control* is to recognize when the accumulated experimental data shows that a controller fails to achieve the preferred stability and performance objectives. If a destabilizing controller happens to be the currently active one, then adaptive control should eventually switch it out of the loop, and replace it with an optimal, stabilizing one. An optimal controller is one that optimizes the controller ordering criterion ("cost function") given the currently available evidence. This perspective renders the adaptive control problem in a form of a standard constrained optimization. A concept similar to this feasibility notion can be found in [67].

To address the emerging need for robustness for larger uncertainties or achieve tighter performance specifications, several recent important advances have emerged, such as [57] and *multi-model controller switching* formulations of the adaptive control problem, *e.g.*, supervisory based control design in [41,46,72,77] or data-driven unfalsified adaptive control methods of [15, 18, 90, 99] (based on criteria of falsifiability [82, 112, 117]) which exploit evidence in the plant output data to switch a controller out of the loop when the evidence proves that the controller is failing to achieve the stated goal. In both cases, the outer supervisory loop introduced to the baseline adaptive system allows fast 'discontinuous' adaptation in highly uncertain nonlinear systems, and thus leads to improved performance and overcomes some limitations of classical adaptive control. These formulations have led to improved optimization based adaptive control theories and, most importantly, significantly weaker assumptions of prior knowledge. Both indirect [72,77,118,119] and direct [37,67,90] switching methods have been proposed for the adaptive supervisory loop. Recently, performance issues in switching adaptive control have been addressed in Robust Multiple Model Adaptive Control schemes (RMMAC) [8]. Various extensions of the fundamental unfalsified control ideas that form the basis for the safe switching control have been reported recently. These include, for example, stability in the noisy environment [13], switching supervisory control [12]; unfalsified control of photovoltaic energy storage system [24]; unfalsified control experiments on dual rotary fourth-order motion system with real hardware [96]; ellipsoidal unfalsified control [103], [104], [105], [106], [107]; application of unfalsified control to a stirred tank reactor; and [27], [30], [50], [55], [66], [93]. The reference [115] asserts that unfalsified control is possibly better alternative for "fault tolerant control (FTC)" flight control reconfiguration.

The results of this study build on the result of Morse *et al.* [72,75] and Hespanha *et al.* [45,46,47]. The theoretical ground in [95] is widened by allowing the class of candidate controllers to be infinite so as to allow consideration of continuously parameterized adaptive controllers, in addition to finite sets of controllers. Under some mild additional assumptions on the cost function (designer based, not plant-dependent), stability of the closed loop switched system is assured, as well as the convergence to a stabilizing controller in finitely many steps. An academic example shows that, when there is a mismatch between the true plant and the assumptions on the plant ('priors'), a wrong ordering of the controllers can in some cases give preference to destabilizing controllers. This phenomenon is called *model mismatch instability*. However, prior knowledge and plant models, when they can be found, can be incorporated into the design of the candidate controllers, which could be used together with the safe switching algorithm.

In certain ways, the paradigm of adaptive control problem cast as a constrained optimization problem bears similarities with the ideas found in machine learning algorithms [70,97]. As a direct, data-driven switching adaptive method, it is more similar to the reinforcement learning algorithms than supervised/unsupervised learning. In reinforcement learning, the algorithm learns a policy of how to act given an observation of the world. Every action has some impact in the environment, and the environment provides feedback that guides the learning algorithm.

## 2.2 Preliminaries

The system under consideration in this book is the adaptive system $\Sigma : \mathfrak{L}_{2e} \longrightarrow \mathfrak{L}_{2e}$ shown in Figure 2.1, where $u$ and $y$ are the plant input and output vector signals, and $\mathfrak{L}_{2e}$ is the linear vector space (extended Lebesgue space) of functions $x(t)$ whose $\mathfrak{L}_{2e}$ norm, defined as $||x||_\tau \triangleq (\int_0^\tau x(t)^T x(t)dt)^{1/2}$, exists for any finite $\tau$. For any $\tau \in \mathbf{T} = \mathbb{R}_+$, a *truncation operator* $P_\tau$ is a linear projection operator that truncates the signal at $t = \tau$. The symbol $x_\tau$ will be used for the truncated signal $P_\tau x$ [88]. The adaptive controller switches the currently active controller $\hat{K}_t$ at times $t_k, k = 1,2,\ldots$ with $t_k < t_{k+1}, \forall k$. For brevity, we also denote $K_k = \hat{K}_{t_k}$ the controller switched in the loop during the time interval $t \in [t_k, t_{k+1})$. If finite, the total number of switches is denoted by $N$, so that the final switching time is $t_N$ and the final controller is $K_N$.

We define the set $\mathbf{Z} = \text{Graph}\{\mathscr{P}\} \triangleq \{z = (u,y) \big| y = \mathscr{P}u\}$ where $\mathscr{P}$ is an unknown plant. Unknown disturbances and noises $(d,n)$ may also affect the plant relation $\mathscr{P}$. Let $z_d = (y_d, u_d) \in \mathbf{Z}$ represent the output data signals measured in one experiment, defined on the time interval $[0,\infty)$.

We consider a possibly infinite set $\mathbf{K}$ (*e.g.*, a set containing a continuum) of the candidate controllers. The finite controller set case is included as a special case. The parametrization of $\mathbf{K}$, denoted $\Theta_\mathbf{K}$, will initially be taken to be a subset of $\mathbb{R}^n$; the treatment of the infinite dimensional spaces will be discussed in the Remark 2.5.

We now recall some familiar definitions from the stability theory. A function $\phi : \mathbb{R}_+ \to \mathbb{R}_+$ belongs in *class* $\mathscr{K}$ ($\phi \in \mathscr{K}$) if $\phi$ is continuous, strictly increasing and $\phi(0) = 0$. The $\mathfrak{L}_2$-*norm* of a truncated signal $P_\tau z$ is given as $||z||_\tau = \sqrt{\int_0^\tau z(t)^T z(t)dt}$. The Euclidean norm of the parameterization $\theta_K \in \mathbb{R}^n$ of the controller $K$ is denoted $||\theta_K||$. A functional $f : \mathbb{R}^n \to \mathbb{R}$ is said to be *coercive* [14] if $\lim f(x) = \infty$ when $||x|| \to \infty, x \in \mathbb{R}^n$.

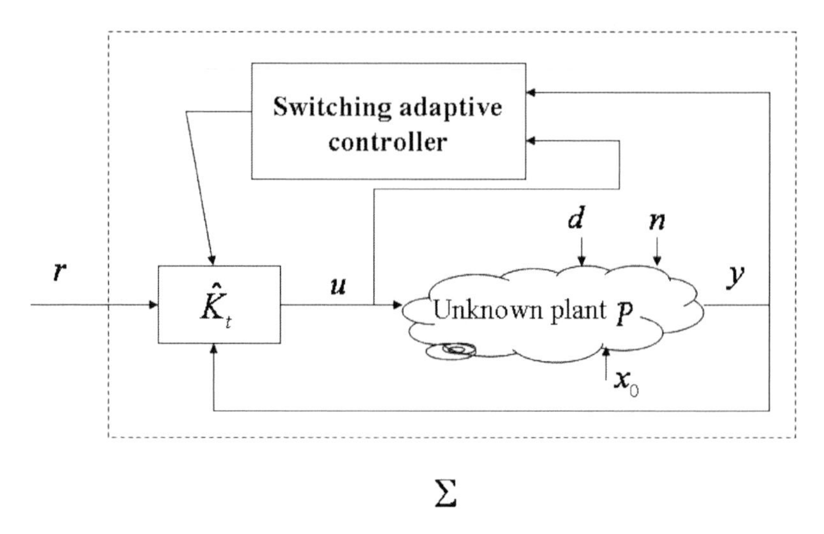


**Fig. 2.1** Switching adaptive control system $\Sigma(\hat{K}_t, P)$

**Definition 2.1.** A system $\Sigma : \mathfrak{L}_{2e} \longrightarrow \mathfrak{L}_{2e}$ with input $w$ and output $z$ is said to be *stable* if for every such input $w \in \mathfrak{L}_{2e}$ there exist constants $\beta, \alpha \geq 0$ such that

$$||z||_\tau < \beta ||w||_\tau + \alpha, \forall \tau > 0\,. \tag{2.1}$$

Otherwise, $\Sigma$ is said to be *unstable*. Furthermore, if (2.1) holds with a single pair $\beta, \alpha \geq 0$ for all $w \in \mathfrak{L}_{2e}$, then the system $\Sigma$ is said to be *finite-gain stable*, in which case the *gain* of $\Sigma$ is the least such $\beta$.

*Remark 2.1*. In general, $\alpha$ can depend on the initial state.

Specializing to the system in Figure 2.1, and (without loss of generality) disregarding $d, n, x_0$, stability of the closed loop system $\Sigma$ means that for every $r \in \mathfrak{L}_{2e}$, there exist $\beta, \alpha \geq 0$ such that $||[y,u]||_\tau \leq \beta ||r||_\tau + \alpha$.

**Definition 2.2.** The system $\Sigma$ is said to be *incrementally stable* if, for every pair of inputs $w_1, w_2$ and outputs $z_1 = \Sigma w_1, z_2 = \Sigma w_2$, there exist constants $\tilde{\beta}, \tilde{\alpha} \geq 0$ such that

$$||z_2 - z_1||_\tau < \tilde{\beta} ||w_2 - w_1||_\tau + \tilde{\alpha}, \forall \tau > 0 \tag{2.2}$$

and the *incremental gain* of $\Sigma$, when it exists, is the least $\tilde{\beta}$ satisfying (2.2) for some $\tilde{\alpha}$ and all $w_1, w_2 \in \mathfrak{L}_{2e}$.

**Definition 2.3.** The adaptive control problem is said to be *feasible* if a candidate controller set $\mathbf{K}$ contains at least one controller such that the system $\Sigma(K, \mathscr{P})$ is stable. A controller $K \in \mathbf{K}$ is said to be a *feasible controller* if the system $\Sigma(K, \mathscr{P})$ is stable.

*Safe adaptive control problem goal* is then formulated as finding an asymptotically optimal, stabilizing controller, given the feasibility of the adaptive control problem. Under this condition, safe adaptive control should recognize a destabilizing controller currently existing in the loop, and replace it with an as yet unfalsified controller. Hence, we have the following (and the only) assumption on the plant that we will use.

**Assumption 2.1.** The adaptive control problem is feasible.

A similar notion of the safe adaptive control appears in [4], where Anderson *et al.* define the safe adaptive switching control as one that always yields stable frozen closed loop; the solution to this problem is achieved using the $\nu$-gap metric. Limitations of the $\nu$-gap metric are discussed in [49].

Prevention from inserting a destabilizing controller in the loop is not assured (since the adaptive switching system cannot identify with certainty a destabilizing controller beforehand, based on the past data), but if such a controller is selected, it will quickly be switched out as soon as the unstable modes are excited. Under the feasibility condition, an unfalsified controller will always be found and placed in the loop. Whether the optimal, robustly stabilizing controller will eventually be found and connected, depends on whether the unstable modes are sufficiently excited.

It follows from the above definition of the adaptive control goal that feasibility is a necessary condition for the existence of a particular $K \in \mathbf{K}$ that robustly solves the safe adaptive control problem. The results in this monograph show that feasibility is also a sufficient condition to design a robustly stable adaptive system that converges to a $K \in \mathbf{K}$, even when it is not known *a priori* which controllers $K$ in the set $\mathbf{K}$ are stabilizing.

**Definition 2.4.** Stability of the system $\Sigma : w \mapsto z$ is said to be *unfalsified* by the data $(w, z)$ if there exist $\beta, \alpha \geq 0$ such that (2.1) holds; otherwise, we say that stability of the system $\Sigma$ is *falsified* by $(w, z)$.

Unfalsified stability is determined from (2.1) based on the data from one experiment for one input, while 'stability' requires additionally that (2.1) hold for the data from every possible input.

Any adaptive control scheme has a cost index inherently tied to it, which orders controllers based on evidence found in data. This index is taken here to be a *cost functional* $V(K, z, t)$, defined as a causal-in-time mapping:

$$V : \mathbf{K} \times \mathbf{Z} \times \mathbf{T} \to \mathbb{R}_{+} \cup \{\infty\} .$$

An example of the cost function according to the above definition, which satisfies the desired properties introduced later in the text, is given in Section 2.4.

The switched system comprised of the plant $\mathscr{P}$ and the currently active controller $\hat{K}_t$, where $\hat{K}_t = K_k$, $K_k \in \mathbf{K}$ is denoted $\Sigma(\hat{K}_t, \mathscr{P})$ (Figure 2.1). For the switched system $\Sigma(\hat{K}_t, \mathscr{P})$ in Figure 2.1, the *true cost* $V_{true} : \mathbf{K} \to \mathbb{R}_{+} \cup \{\infty\}$ is defined as $V_{true}(K) = \sup_{z \in \mathbf{Z}, \tau \in \mathbf{T}} V(K, z, \tau)$, where $\mathbf{Z} = \text{Graph}\{\mathscr{P}\}$.

**Definition 2.5.** Given the pair $(V, \mathbf{K})$, a controller $\overline{K} \in \mathbf{K}$ is said to be falsified at time $\tau$ by the past measurement $z_\tau$ if $V(\overline{K}, z, \tau) > \min_{K \in \mathbf{K}} V(K, z, \tau) + \varepsilon$. Otherwise it is said to be unfalsified at time $\tau$ by $z_\tau$.

Then, a *robust optimal controller* $K_{RSP}$ is one that stabilizes (in the sense of the Definition 2.1) the given plant and minimizes the true cost $V_{true}$.

Therefore, $K_{RSP} = \arg\min_{K \in \mathbf{K}} V_{true}(K)$ (and is not necessarily unique). Owing to the feasibility assumption, at least one such $K_{RSP}$ exists, and $V_{true}(K_{RSP}) < \infty$.

**Definition 2.6.** [90] For every $K \in \mathbf{K}$, a *fictitious reference signal* $\tilde{r}_K(z_d, \tau)$ is defined to be an element of the set

$$\tilde{R}(K, z_d, \tau) \triangleq \{r | K \begin{bmatrix} r \\ y \end{bmatrix} = u,\ P_\tau z_d = \begin{bmatrix} P_\tau u \\ P_\tau y \end{bmatrix}\}. \tag{2.3}$$

In other words, $\tilde{r}_K(z_d, \tau)$ is a hypothetical reference signal that would have exactly reproduced the measured data $z_d$ had the controller $K$ been in the loop for the entire time period over which the data $z_d$ was collected.

**Definition 2.7.** When for each $z_d$ and $\tau$ there is a unique $\tilde{r} \in \tilde{R}(K, z_d, \tau)$, then we say $K$ is *causally left invertible (CLI)* and we denote by $\mathfrak{K}_{CLI}$ the induced causal map $z_d \mapsto \tilde{r}$. The causal left inverse $\mathfrak{K}_{CLI}$ is called the *fictitious reference signal generator* (FRS) for the controller $K$. When $\mathfrak{K}_{CLI}$ is incrementally stable, $K$ is called *stably causally left invertible controller* (SCLI).

**Definition 2.8.** Let $r$ denote the input and $z_d = \Sigma(\hat{K}_t, \mathscr{P})r$ denote the resulting plant data collected with $\hat{K}_t$ as the current controller. Consider the adaptive control system $\Sigma(\hat{K}_t, \mathscr{P})$ of Figure 2.1 with input $r$ and output $z_d$. The pair $(V, \mathbf{K})$ is said to be *cost detectable* if, without any assumption on the plant $P$ and for every $\hat{K}_t \in \mathbf{K}$ with finitely many switching times, the following statements are equivalent:

- $V(K_N, z_d, t)$ is bounded as $t$ increases to infinity.
- Stability of the system $\Sigma(\hat{K}_t, \mathscr{P})$ is unfalsified by the input-output pair $(r, z_d)$.

*Remark 2.2.* With cost detectability satisfied, we can use the cost $V(K, z, t)$ to reliably detect any instability exhibited by the adaptive system, *even when initially the plant is completely unknown.*

*Remark 2.3.* Cost detectability is different from the plant *detectability*. Cost detectability is determined from the knowledge of the cost function and candidate controllers, without reference to the plant. In [45], a problem similar to ours is approached using the following assumptions: 1) the plant itself is detectable, and 2) the candidate plant models are stabilized by the corresponding candidate controllers. The difference between the approach in [42, 45] and this study lies in the definition of cost detectability introduced in this study, which is the property of the cost function/candidate-controller-set pair, but is independent of the plant.

In the following, we use the notation $\mathscr{V} \doteq \{V_{z,t} : z \in \mathbf{Z}, t \in \mathbf{T}\} : \mathbf{K} \to \mathbb{R}_+$ for a family of functionals with the common domain $\mathbf{K}$, with $V_{z,t}(K) \doteq V(K,z,t)$. Let $\mathbf{L} \doteq \{K \in \mathbf{K} | V_{z,t_0}(K) \leq V_{true}(K_{RSP}), V \in \mathscr{V}\}$ denote the level set in the controller space corresponding to the cost at the first switching time instant. With the family of functionals $\mathscr{V}$ with a common domain $\mathbf{K}$, a *restriction* to the set $\mathbf{L} \subseteq \mathbf{K}$ is associated, defined as a family of functionals $\mathscr{W} \doteq \{W_{z,t}(K) : z \in \mathbf{Z}, t \in \mathbf{T}\}$ with a common domain $\mathbf{L}$. Thus, $W_{z,t}(K)$ is identical to $V_{z,t}(K)$ on $\mathbf{L}$, and is equal to $V_{true}(K_{RSP}$ outside $\mathbf{L}$.

Consider now the cost minimization hysteresis switching algorithm reported in [75], together with the cost functional $V(K,z,t)$. The algorithm returns, at each $t$, a controller $\hat{K}_t$ which is the active controller in the loop:

---

**Algorithm 2.1.** The $\varepsilon$-hysteresis switching algorithm [75]. At each time instant $t$, the currently active controller is defined as:

$$\hat{K}_t = \arg\min_{K \in \mathbf{K}} \{V(K,z,t) - \varepsilon \delta_{K\hat{K}_{t^-}}\}$$

where $\delta_{ij}$ is the Kronecker's $\delta$, and $t^-$ is the limit of $\tau$ from below as $t \to \tau$.

---

The switch occurs only when the current unfalsified cost related to the currently active controller exceeds the minimum (over $\mathbf{K}$) of the current unfalsified cost by at least $\varepsilon$ (Figure 2.2). The hysteresis step $\varepsilon$ serves to limit the number of switches on any finite time interval to a finite number, and so prevents the possibility of the limit cycle type of instability. It also ensures a non-zero dwell time between switches

The hysteresis switching lemma of [75] implies that a switched sequence of controllers $\hat{K}_{t_k}$ $(k = 1,2,\ldots)$, which minimize (over $\mathbf{K}$) the current unfalsified cost $V(K,z,t)$ at each switch-time $t_k$, will also stabilize the plant if the cost related to each fixed controller $K$ has the following properties: first, it is a monotone increasing function of time and second, and it is uniformly bounded above if and only if $K$ is stabilizing. But, these properties were demonstrated for the cost functions $V(K,z,t)$ in [75] only by introducing prior assumptions on the plant, thereby also introducing the possibility of model mismatch instability.

**Definition 2.9.** [110] Let $\mathbf{S}$ be a topological space. A family $\mathscr{F} \doteq \{f_\alpha : \alpha \in \mathbf{A}\}$ of complex functionals with a common domain $\mathbf{S}$ is said to be *equicontinuous at a point* $x \in \mathbf{S}$ if for every $\varepsilon > 0$ there exists an open neighborhood $N(x)$ such that $\forall y \in N(x)$, $\forall \alpha \in \mathbf{A}$, $|f_\alpha(x) - f_\alpha(y)| < \varepsilon$. The family is said to be *equicontinuous* on $\mathbf{S}$ if it is equicontinuous at each $x \in \mathbf{S}$. $\mathscr{F}$ is said to be *uniformly equicontinuous* on $\mathbf{S}$ if $\forall \varepsilon > 0$, $\exists \delta = \delta(\varepsilon) > 0$ such that $\forall x,y \in \mathbf{S}$, $\forall \alpha \in \mathbf{A}$, $y \in N_\delta(x) \Rightarrow |f_\alpha(x) - f_\alpha(y)| < \varepsilon$, where $N_\delta$ denotes an open neighborhood of size $\delta$.

In a metric space $\mathbf{S}$ with a metric $d_{\mathbf{S}}$, uniform equicontinuity means that $\forall x,y \in \mathbf{S}$, $\forall \alpha \in \mathbf{A}$, $d_{\mathbf{S}}(x,y) < \delta \Rightarrow |f_\alpha(x) - f_\alpha(y)| < \varepsilon$.

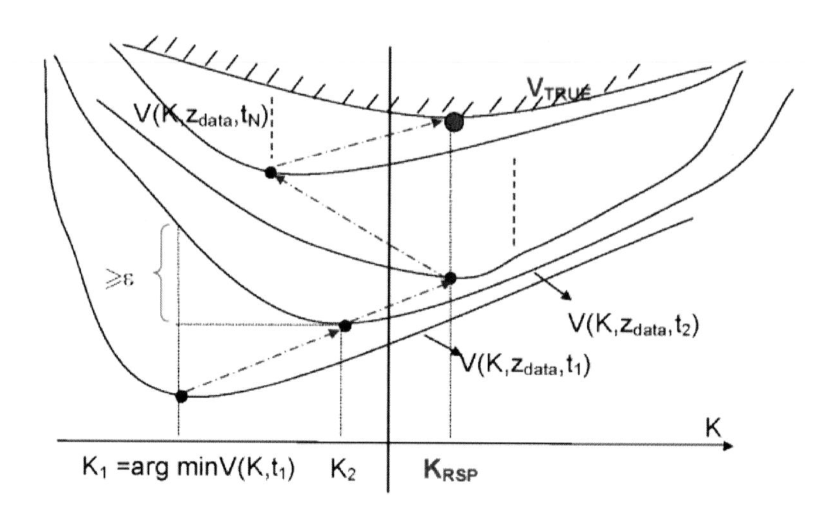


**Fig. 2.2** Cost vs. control gain time snapshots

**Lemma 2.1.** *If $(S,d)$ is a compact metric space, then any family $\mathscr{F} \doteq \{f_\alpha : \alpha \in A\}$ that is equicontinuous on* **S** *is uniformly equicontinuous on* **S**.

*Proof.* The proof is a simple extension of the theorem that asserts uniform continuity property of a continuous mapping from a compact metric space to another metric space [86]. Let $\varepsilon > 0$ be given. Since $\mathscr{F} \doteq \{f_\alpha : \alpha \in \mathbf{A}\}$ is equicontinuous on **S**, we can associate to each point $p \in \mathbf{S}$ a positive number $\phi(p)$ such that $q \in S, d_{\mathbf{S}}(p,q) < \phi(p) \Rightarrow |f_\alpha(p) - f_\alpha(q)| < \frac{\varepsilon}{2}$, for all $\alpha \in \mathbf{A}$. Let $J(p) \doteq \{q \in \mathbf{S} \mid d_{\mathbf{S}}(p,q) < \frac{1}{2}\phi(p)\}$. Since $p \in J(p)$, the collection of all sets $J(p), p \in \mathbf{S}$ is an open cover of **S**; and since **S** is compact, there is a finite set of points $p_1, \ldots, p_n \in \mathbf{S}$ such that $\mathbf{S} \subseteq \cup_{i=1,n} J(p_i)$. Let us set $\delta = \frac{1}{2}\min_{1 \le i \le n}[\phi(p_i)]$. Then $\delta > 0$, because a minimum of a finite set of positive numbers is positive, as opposed to the *inf* of an infinite set of positive numbers which may be 0. Now let $p,q \in \mathbf{S}$, $d_{\mathbf{S}}(p,q) < \delta$. Then, $p \in J(p_m), m \in 1,\ldots,n$. Hence, $d_{\mathbf{S}}(p,p_m) < \frac{1}{2}\phi(p_m)$ and $d_{\mathbf{S}}(q,p_m) < d_{\mathbf{S}}(q,p) + d_{\mathbf{S}}(p,p_m) < \delta + \frac{1}{2}\phi(p_m) < \phi(p_m)$. Thus, $|f_\alpha(p) - f_\alpha(q)| < |f_\alpha(p) - f_\alpha(p_m)| + |f_\alpha(p_m) - f_\alpha(q)| < \frac{\varepsilon}{4} + \frac{\varepsilon}{2} < \varepsilon \; \forall \alpha \in \mathbf{A}$. Since $\delta$ holds for all $p,q \in \mathbf{S}$ and all $\alpha \in \mathbf{A}$, $\mathscr{F}$ is uniformly equicontinuous. □

## 2.3 Results

The main results on stability and finiteness of switches are developed in the sequel.

**Lemma 2.2.** *Consider the feedback adaptive control system $\Sigma$ in Figure 2.1 with input $r$ and output $z_d = (u,y)$, together with the hysteresis switching Algorithm 2.1 (generality is not lost if $r$ is taken instead of the input $w = [r\ d\ n\ x_0]$). Suppose there are finitely many switches. If the adaptive control problem is feasible (Definition 2.3), candidate controllers are SCLI, and the following properties are satisfied:*

- *$(V,\mathbf{K})$ is cost detectable (Definition 2.8)*
- *$V$ is monotone increasing in time*

*then the final switched controller is stabilizing. Moreover, the system response $z$ with the final controller satisfies the performance inequality*

$$V(K_N,z,t) \leq V_{true}(K_{RSP}) + \varepsilon \;\; \forall t \, .$$

*Proof.* It suffices to consider the final controller $K_N$. Denote the final switching time instant $t_N$. Then, by the definition of $V_{true}(K_N)$, and feasibility of the control problem (Definition 2.3), it follows that for all $t \geq t_N$,

$$\begin{aligned} V(K_N,z_d,t) &< \varepsilon + \min_K V(K,z_d,t) \\ &< \varepsilon + V_{true}(K_{RSP}) \; < \infty. \end{aligned} \tag{2.4}$$

Further, by monotonicity in $t$ of $V(K,z,t)$, it follows that (2.4) holds for all $t \in \mathbf{T}$. Owing to the cost detectability, stability of $\Sigma$ with $K_N$ is not falsified by $z_d$, that is, there exist constants $\beta, \alpha \geq 0$ corresponding to the given $\tilde{r}_{K_N}$ such that

$$||z_d||_t < \beta ||\tilde{r}_{K_N}||_\tau + \alpha, \; \forall t > 0 \, . \tag{2.5}$$

According to Lemma A.1 in Appendix Appendix A, there exist $\beta_1, \alpha_1 \geq 0$ such that $||\tilde{r}_{K_N}||_t < \beta_1 ||r||_t + \alpha_1, \forall t > 0$. This, along with (2.5) implies $||z_d||_t < \beta_2 ||r||_t + \alpha_2, \forall t > 0$, for some $\beta_2, \alpha_2 \geq 0$. □

**Lemma 2.3.** *Let $f : \mathbb{R}^n \to \mathbb{R}$ be a continuous and coercive function on $\mathbb{R}^n$. Then for any scalar $\alpha \in \mathbb{R}$, the level set $L(\alpha) \doteq \{x \in \mathbb{R}^n \mid f(x) \leq \alpha\}$ is compact.*

*Proof.* Since $L(\alpha) \subset \mathbb{R}^n$, we show that $L(\alpha)$ is closed and bounded: Let $\{x_m\} \subseteq L(\alpha)$ be a convergent sequence, and $\bar{x} \doteq \lim_{m\to\infty} x_m$. Since $f$ is continuous, $f(\bar{x}) = \lim_{m\to\infty} f(x_m)$. Also, $f(x_m) \leq \alpha, \forall m \in \mathbb{N}$. Then, $f(\bar{x}) = \lim_{m\to\infty} f(x_m) \leq \lim_{m\to\infty} \alpha = \alpha$, so $\bar{x} \in L(\alpha)$. Hence, $L(\alpha)$ is closed. To show that is $L(\alpha)$ is bounded, proceed by contradiction. Assume that $L(\alpha)$ is not bounded; then there exists a sequence $\{y_m\} \subseteq L(\alpha)$ such that $\lim_{m\to\infty} ||y_m|| = \infty$. Since $f$ is coercive, $\lim_{m\to\infty} f(y_m) = \infty$; in particular, $\exists N \in \mathbb{N}$ such that $\forall k \geq N \;\, f(y_k) > \alpha$, for any fixed $\alpha \in \mathbb{R}$. Then, $\{y_m\} \not\subseteq L(\alpha)$, which contradicts the above assumption. Thus, $L(\alpha)$ is closed and bounded in $\mathbb{R}^n$, therefore compact. □

**Lemma 2.4.** *Consider the feedback adaptive control system in Figure 2.1, together with the switching Algorithm 2.1. If the adaptive control problem is feasible (Definition 2.3), and the associated cost functional/controller set pair $(V,\mathbf{K})$ is cost detectable, $V$ is monotone increasing in time and, in addition,*

- *For all $\tau \in \mathbf{T}, z \in \mathbf{Z}$, the cost functional $V(K,z,t)$ is* coercive *on $\mathbf{K} \subseteq \mathbb{R}^n$ (i.e. $\lim_{||K||\to\infty} V(K,z,\tau) = \infty$), and*
- *The family $\mathscr{W} \doteq \{W_{z,t}(K) : z \in \mathbf{Z}, t \in \mathbf{T}\}$ of restricted cost functionals with a common domain $\mathbf{L} \doteq \{K \in \mathbf{K} | V_{z,t_0}(K) \leq V_{true}(K_{RSP}), V \in \mathscr{V}\}$ is equicontinuous on $\mathbf{L}$,*

*then the number of switches is uniformly bounded above for all $z \in \mathbf{Z}$ by some $\bar{N} \in \mathbb{N}$.*

*Proof.* Our proof is similar to the convergence lemmas of [46, 75]. A graphical representation of the switching process, giving insight to the derivation presented below, is shown in Figure 2.3.

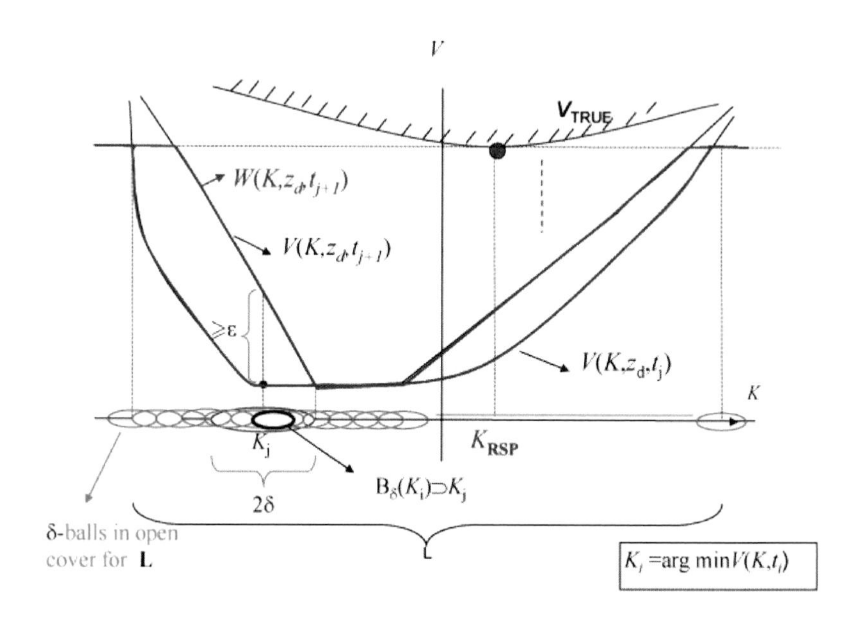


**Fig. 2.3** Derivation of the upper bound on switches in the continuum **K** case.

According to Lemma 2.3, the level set **L** is compact. Then, the family $\mathscr{W} \doteq \{W_{z,t}(K) : z \in \mathbf{Z}, t \in \mathbf{T}\}$ is uniformly equicontinuous on **L** (Lemma 2.1), *i.e.*, for a hysteresis step $\varepsilon$, $\exists \delta > 0$ such that for all $z \in \mathbf{Z}$, $t \in \mathbf{T}$, $K_1, K_2 \in \mathbf{L}$, $||\theta_{K_1} - \theta_{K_2}|| < 2\delta \Rightarrow |W_{z,t}(K_1) - W_{z,t}(K_2)| < \varepsilon$ (*i.e.* $\delta = \delta(\varepsilon)$ is common to all $K \in \mathbf{L}$ and all $z \in \mathbf{Z}, t \in \mathbf{T}$). Since **L** is compact, there exists a finite open cover $\mathscr{C}_N = \{B_\delta(K_i)\}_{i=1}^N$ , with $K_i \in \mathbb{R}^n, i = 1, \ldots, N$ such that $\mathbf{L} \subset \cup_{i=1}^N B_\delta(K_i)$, where $N$ depends on the chosen hysteresis step $\varepsilon$ (this is a direct consequence of the definition of a compact set). Let $\hat{K}_{t_j}$ be the controller switched into the loop at the time $t_j$, and the corresponding minimum cost achieved is $\tilde{V} \doteq \min_{K \in \mathbf{K}} V(K, z, t_j)$. Consider that at the time $t_{j+1} > t_j$ a switch occurs at the same cost level $\tilde{V}$, i.e. $\tilde{V} = \min_{K \in \mathbf{K}} V(K, z, t_{j+1})$ where $V(\hat{K}_{t_j}, z, t_{j+1}) > \min_{K \in \mathbf{K}} V(K, z, t_{j+1}) + \varepsilon$. Therefore, $\hat{K}_{t_j}$ is falsified, and so are all the controllers $K \in B_{2\delta}(\hat{K}_{t_j})$. Let $I_j$ be the index set of the as yet unfalsified $\delta$-balls of controllers at the time $t_j$. Since $\hat{K}_{t_j} \in B_\delta(K_i)$, for some $i \in \bar{I} \subset I_j$ ($\bar{I}$ is not necessarily a singleton as $\hat{K}_{t_j}$ may belong to more than one balls $B_\delta(K_i)$, but it suffices for the proof that there is at least one such index $i$), also falsified are all the controllers $K \in B_\delta(K_i) \supset \hat{K}_{t_j}$, so that $I_{j+1} = I_j \setminus \{i\}$. Thus, $I_j$ is updated according to the following algorithm, in which $j$ denotes the index of the switching time $t_j$:

---

**Algorithm 2.2.** Unfalsified index set algorithm:

1. Initialize: Let $j = 0$, $I_0 = \{1,\ldots,N\}$
2. $j \leftarrow j+1$. If $I_{j-1} = \varnothing$: Set $I_j = \{1,\ldots,N\}$ // Optimal cost increases
   Else $I_j = I_{j-1} \setminus \{i\}$,where $i \in I_{j-1}$ is such that $B_\delta(K_i) \supset \hat{K}_{t_{j-1}}$
3. go to (2)

---

Therefore, the number of possible switches to a single cost level is upper-bounded by $N$, the number of $\delta$-balls in the cover of $\mathbf{L}$. The next switch (the very first after the $N^{th}$ one), if any, must occur to a cost level higher than $\tilde{V}$, because of the monotonicity of $V$. But then, according to Algorithm 2.1, $|V(\tilde{K}_{t_{j+N+1}}, z, t_{j+N+1}) - \tilde{V}| > \varepsilon$, with $d(\tilde{K}_{t_{j+N+1}}, \tilde{K}_{t_k}) < 2\delta$, $j \leq k \leq j+N$ and $V(\tilde{K}_{t_k}, z, t_k) = \tilde{V}$. Combining the two bounds, the overall number of switches is upper-bounded by:

$$\bar{N} \doteq N \frac{V_{true}(K_{RSP}) - \min_{K \in \mathbf{K}} V(K,z,0)}{\varepsilon} . \qquad \square$$

Equicontinuity assures that the cost functionals in the said family have associated $\delta$-balls of finite, non-zero radii, which is used to upper bound the number of switches. If $\Theta_{\mathbf{K}} \subseteq \mathbb{R}^n$ holds, then the set $\mathbf{L}$ is compact; otherwise an additional requirement that the set $\mathbf{K}$ is compact is needed. The finite controller set case is obtained as a special case of the Lemma 2.4, with $N$ being the number of candidate controllers instead of the number of $\delta$-balls in the cover of $\mathbf{L}$. The main result follows.

**Theorem 2.1.** *Consider the feedback adaptive control system $\Sigma$ in Figure 2.1, together with the hysteresis switching Algorithm 2.1. Suppose that the adaptive control problem is feasible, the associated cost functional $V(K,z,t)$ is monotone in time, the pair $(V,\mathbf{K})$ is cost detectable, candidate controllers are SCLI, and the conditions of Lemma 2.4 hold. Then, the switched closed loop system is stable, according to Definition 2.1. In addition, for each $z$, the system converges after finitely many switches to the controller $K_N$ that satisfies the performance inequality*

$$V(K_N, z, t) \leq V_{true}(K_{RSP}) + \varepsilon \ \ \forall t . \tag{2.6}$$

*Proof.* Invoking Lemma 2.4 proves that there are finitely many switches. Then, Lemma 2.2 shows that the adaptive controller stabilizes, according to Definition 2.1, and that (3.75) holds. $\square$

*Remark 2.4.* Note that, due to the coerciveness of $V$, $\min_{K \in \mathbf{K}} V(K,z,0)$ is bounded below (by a non-negative number, if the range of $V$ is a subset of $\mathbb{R}_+$), for all $z \in \mathbf{Z}$.

*Remark 2.5.* The parametrization of the candidate controller set can be more general than $\Theta_{\mathbf{K}} \subseteq \mathbb{R}^n$; in fact, it can belong to an arbitrary infinite dimensional space; however, $\mathbf{K}$ has to be compact in that case, to ensure uniform equi-continuity property.

Note that the switching ceases after finitely many steps for all $z \in \mathbf{Z}$. If the system input is sufficiently rich so as to increase the cost more than $\varepsilon$ above the level at the time of the latest switch, then a switch to a new controller that minimizes the current cost will eventually occur at some later time. The values of these cost minima at any time are monotone increasing and bounded above by $V_{true}(K_{RSP})$. Thus, sufficient richness of the system input (external reference signal, disturbance or noise signals) will affect the cost to approach $V_{true}(K_{RSP}) \pm \varepsilon$.

*Remark 2.6.* The minimization of the cost functional over the infinite set $\mathbf{K}$ is tractable if the compact set $\mathbf{K}$ can be represented as a finite union of convex sets, *i.e.*, the cost minimization is a convex programming problem.

## 2.4 Cost Function Example

An example of the cost function and the conditions under which it ensures stability and finiteness of switches according to Theorem 1 may be constructed as follows. Consider (a not necessarily zero-input zero-output) system $\Sigma : \mathfrak{L}_{2e} \rightarrow \mathfrak{L}_{2e}$ in Figure 2.1. Choose as a cost functional:

$$V(K,z,t) = \max_{\tau \leq t} \frac{||y||_\tau^2 + ||u||_\tau^2}{||\tilde{r}_K||_\tau^2 + \alpha} + \beta + \gamma ||K||^2 \tag{2.7}$$

where $\alpha$, $\beta$, and $\gamma$ are arbitrary positive numbers. The constant $\alpha$ is used to prevent division by zero when $\tilde{r} = y = u = 0$ (unless $\Sigma$ has zero-input zero-output property), $\beta$ ensures $V > 0$ even when $||K|| \equiv 0$, and $\gamma > 0$ ensures coerciveness of $V(K,z,t)$. Alternatively, in order to avoid the restriction to the minimum phase (SCLI) controllers (which would assure causality and incremental stability of the map $[u,y] \rightarrow \tilde{r}$), the denominator of (2.7) can contain $\tilde{v}_K$ instead of $\tilde{r}_K$ [26], [65], where $\tilde{v}_K$ is defined via the matrix fraction description (MFD) form of the controller $K$, as $K = D_K^{-1} N_K$ and $\tilde{v}_K(t) = (-N_K)(-y(t)) + D_K u(t)$ (Appendix Appendix B), where the relevant signals are shown in Figure 2.4:

$$V(K,z,t) = \max_{\tau \leq t} \frac{||y||_\tau^2 + ||u||_\tau^2}{||\tilde{v}_K||_\tau^2 + \alpha} + \beta + \gamma ||K||^2 \,. \tag{2.8}$$

Both (2.7) and (2.8) satisfy the required properties of Theorem 1, *i.e.*, monotonicity in time, coerciveness on $\mathbf{K}$, equicontinuity of the restricted cost family $\mathscr{W}$, and cost detectability. The first two properties are evident by inspection of (2.7) and (2.8). The justification for the last two properties is as follows.

Since $V(K,z,t)$ in (2.7) and (2.8) is continuous in $K$, then $\mathscr{W}$, defined as the family of the cost functionals $V(K,z,t)$ restricted to the level set $\mathbf{L} \doteq \{K \in \mathbf{K} \mid V(K,z,0) \leq V_{true}\}$, is equicontinuous, since for any $z$ and $t$, $W_{z,t}$ are either equal to $V_{z,t}$, or clamped at $V_{true}$.

**Lemma 2.5.** *Consider the cost functions (2.7) and (2.8) with $\alpha, \beta, \gamma \geq 0$. For $(V, \mathbf{K})$ to be cost detectable, it is sufficient that the candidate controllers in the set $\mathbf{K}$ are SCLI, or that they admit matrix fraction description (MFD) form considered in [65].*

*Proof.* Cost detectability of $(V, \mathbf{K})$ with $V$ in (2.7) follows from the following: 1) the fact that $V(K, z, t)$ is bounded as $t \to \infty$ if and only if stability is unfalsified by the input-output pair $(\tilde{r}_K(t), z)$; 2) SCLI property of the controllers; 3) stability of the mapping $r \mapsto \tilde{r}_{K_N}$ (Lemma A.1 in Appendix Appendix A); and 4) unfalsified stability by the data $(\tilde{r}_{K_N}, z)$ (see Appendix Appendix C). These results can be elaborated further using [65] for the class of non-SCLI controllers and the cost function (2.8), which also ensure 'internal stability' of the adaptive system designed using cost detectable cost-functions of the forms (2.7) or (2.8). □

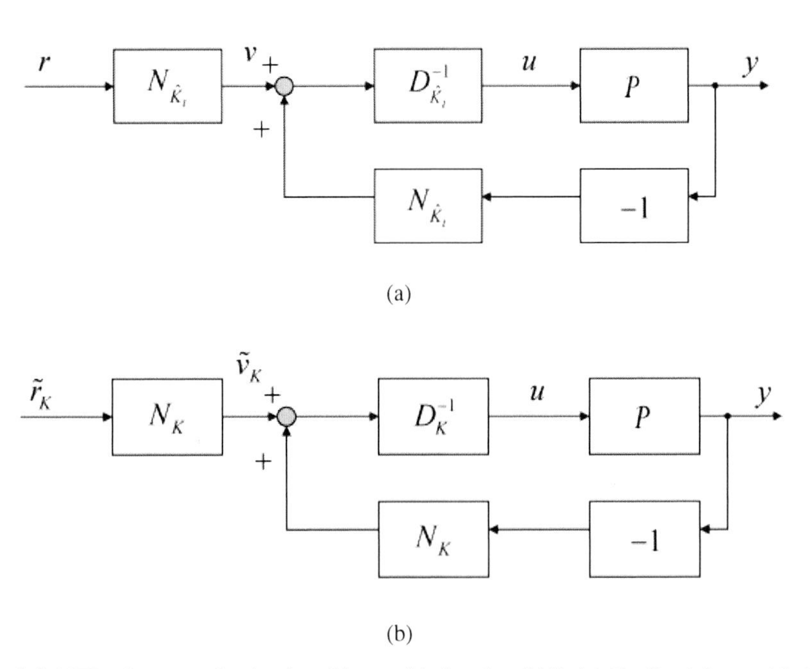


**Fig. 2.4** MFD of a controller in closed loop with the plant [65]: (a) Feedback loop with the current controller $\hat{K}_t$ written in MFD form, (b) Fictitious feedback loop associated with the candidate controller $K$ written in MFD form (in both cases, $(u, y)$ are the actually recorded data)

*Remark 2.7.* Note that the cost function in the preceding example satisfies the $\mathfrak{L}_{2e}$-gain-relatedness property as defined in a previous article by the authors [109].

**Definition 2.10.** ($\mathfrak{L}_{2e}$-gain-related cost). Given a cost/candidate controller-set pair $(V,K)$, we say that the cost $V$ is $\mathfrak{L}_{2e}$-gain-related if for each $z \in \mathfrak{L}_{2e}$ and $K \in \mathbf{K}$,

- $V(K,z,\tau)$ is monotone in $\tau$,
- the fictitious reference signal $\tilde{r}(K,z) \in \mathfrak{L}_{2e}$ exists and,
- for every $K \in \mathbf{K}$ and $z \in \mathfrak{L}_{2e}$, $V(K,z,\tau)$ is bounded as $\tau$ increases to infinity if and only if stability is unfalsified by the input-output pair $(r(K,z)),z)$.

The third condition in Definition 2.10 requires that the cost $V(K,z,\tau)$ be bounded with respect to $\tau$ if and only if $\mathfrak{L}_{2e}$-stability is unfalsified by $(\tilde{r}(K,z),z)$; this is the motivation for the choice of terminology '$\mathfrak{L}_{2e}$-gain-related'. Clearly, cost detectability implies $\mathfrak{L}_{2e}$-gain-relatedness. In fact, $\mathfrak{L}_{2e}$-gain-relatedness is simply cost detectability of $V$ for the special case where $\hat{K}_t \in \mathbf{K}$ is a constant, unswitched non-adaptive controller.

**Lemma 2.6.** *[109] When the candidate controllers have linear time-invariant structure, the sufficient condition for the cost detectability in Lemma 2.5 is also necessary.*

*Proof.* Again, we consider cost functions of the form (2.7) or (2.8); *i.e.*, $\mathfrak{L}_{2e}$-gain-related cost functions. Under this condition, it follows that $\tilde{r}_K(z,\tau)$ is well defined and hence the fictitious reference signal generator $\mathfrak{K}_{CLI}$ exists and is causal. To prove necessity, proceed by contradiction. Suppose that $\mathfrak{K}_{CLI}$ is not stable. Then, the dominant pole of $\mathfrak{K}_{CLI}$ has a non-negative real part, say $\sigma_0 \geq 0$. As cost detectability is a plant-independent property by definition, it must hold for every plant $P$ mapping $\mathscr{L}_{2e} \mapsto \mathfrak{L}_{2e}$. Let us choose $P$ so that $\Sigma(K,P)$ has its dominant closed loop pole at $\sigma_0$. Choose bounded duration inputs $r,s \in \mathfrak{L}_2$ so that the modes of $\mathfrak{K}_{CLI}$ and $\Sigma(K,P)$ associated with the unstable dominant poles with real part $\sigma_0$ are both excited. Then, since the fictitious reference signal $\tilde{r}_K(z,\tau)$ is unstable with the same growth rate $e^{\sigma_0 t}$ as the unstable closed loop response $z(t)$, there exists a constant $\beta$ such that $||z||_t \leq \beta||\tilde{r}_K(z)||_t + \alpha$ holds. Hence, since the cost function satisfies cost detectability, the cost $\lim_{t\to\infty} V(K,z,t)$ is finite. On the other hand, stability of $\Sigma(K,P)$ is falsified by $(r,z)$ which contradicts cost detectability. Therefore, the LTI controller $K$ must be stably causally left invertible (SCLI). □

## 2.5 Specialization to the LTI Plants

It is illuminating to consider how the above results specialize to the case of an LTI plant. In particular, one can derive an explicit bound for the state of the switched system when the final switched controller is the robustly stabilizing and performing controller (denoted $K_*$ in this section for compactness of notation). Let the unknown plant $\mathscr{P}$ in Figure 2.1 be an LTI plant with the control input $u$ and measured output $y$. In addition to a piecewise continuous bounded reference signal $r$, it is assumed that an unknown bounded disturbance $d$ and noise $n$ are acting at the plant input and output, respectively. A set of candidate controllers **K** is considered, which can be an arbitrary infinite set of controllers (of LTI structure in this section). As before, the existence of $K_{RSP} \in \mathbf{K}$ is assumed, such that $K_* = K_{RSP}$ robustly stabilizes $\mathscr{P}$ for any bounded disturbance $d$ and noise $n$ at the plant input and output, respectively.

We will denote the minimal state space representation of the unknown plant $\mathscr{P}$ as $(A_p, B_p, C_p, D_p)$. Let the state-space representation of an individual candidate controller $K$ be denoted as $(A_k, B_k, C_k, D_k)$. Then, the spectral radius of the closed loop state transition matrix $\mathscr{A}_*$ satisfies $\rho(\mathscr{A}_*) < 0$, where

$$\mathscr{A}_* = \left[\begin{array}{c|c} A_{*1} & A_{*2} \\ \hline A_{*3} & A_{*4} \end{array}\right]$$

and

$$\begin{aligned} A_{*1} &\doteq A_p - B_p(I + D_{k^*}D_p)^{-1}D_{k^*}C_p \\ A_{*2} &\doteq B_p(I + D_{k^*}D_p)^{-1}C_{k^*} \end{aligned}$$

and similarly for $A_{*3}$ and $A_{*4}$. As before, $\{t_i\}_{i\in\mathscr{I}}$ is an ordered sequence of the switching time instants for some $\mathscr{I} \subseteq \mathbb{N} \cup \{\infty\}$, with $t_{i+1} > t_i, \forall i \in \mathscr{I}$. The controller switched in the loop at time $t_i, i \in \mathscr{I}$ is denoted $K_i$, whereas $\hat{K}_t$ is the currently active switched controller at time $t$:

$$\hat{K}_t = K_i \ \ on \ \ t \in [t_i, t_{i+1}) \ .$$

Denote the state space realization of $\hat{K}_t$ by $(\hat{A}(t), \hat{B}(t), \hat{C}(t), \hat{D}(t))$, such that, between the switching instants, $\hat{A}(t) = A_i$, $\forall t \in [t_i, t_{i+1})$ (and similarly for other state-space matrices), where $t_i$ denotes the time instant when $K_i$ is switched in the loop.

Let the minimal state space realization for the plant $\mathscr{P}$ be written as:

$$\begin{aligned} \dot{x_p} &= A_p x_p + B_p(u + d) \\ y &= C_p x_p + D_p(u + d) + n \ . \end{aligned}$$

The dynamic equations for the switched controller $\hat{K}_t$ can be written as:

$$\begin{aligned} \dot{\hat{x}} &= \hat{A}(t)\hat{x} + \hat{B}(t)(r - y), \\ u &= \hat{C}(t)\hat{x} + \hat{D}(t)(r - y) \ . \end{aligned}$$

Here, for simplicity of exposition, a one degree-of-freedom (1-DOF) linear controller structure is assumed. The dynamic equations for the piecewise-LTI interconnected system can be written as:

$$\begin{aligned} \dot{x} &= \mathscr{A}(t)x + \mathscr{B}(t)\omega, \\ y &= \mathscr{C}(t)x + \mathscr{D}(t)\omega \end{aligned} \tag{2.9}$$

where $\omega = \begin{bmatrix} d \\ r-n \end{bmatrix}$, $x = \begin{bmatrix} x_p \\ \hat{x} \end{bmatrix}$, and

$$\mathscr{A}(t) = \left[\begin{array}{c|c} \mathscr{A}(t)_1 & \mathscr{A}(t)_2 \\ \hline \mathscr{A}(t)_3 & \mathscr{A}(t)_4 \end{array}\right]$$

with

$$\mathscr{A}(t)_1 \doteq A_p - B_p(I+\hat{D}(t)D_p)^{-1}\hat{D}(t)C_p,$$
$$\mathscr{A}(t)_2 \doteq B_p(I+\hat{D}(t)D_p)^{-1}\hat{C}(t)$$

and

$$\mathscr{B}(t) = \left[\begin{array}{c|c} \mathscr{B}(t)_1 & \mathscr{B}(t)_2 \\ \hline \mathscr{B}(t)_3 & \mathscr{B}(t)_4 \end{array}\right]$$

with

$$\mathscr{B}(t)_1 \doteq B_p(I+\hat{D}(t)D_p)^{-1},$$
$$\mathscr{B}(t)_2 \doteq B_p(I+\hat{D}(t)D_p)^{-1}\hat{D}(t)$$

and similarly for $\mathscr{C}(t)$, $\mathscr{D}(t)$.

Note that $\hat{x}$ is differentiable, if:

- the state of the previously active controller is retained as the initial state of the newly switched controller (due to the requirement of bumpless switching, needed for smooth performance), and
- the states of individual controllers are differentiable in time.

When the final switched controller is the robustly stabilizing and performing controller $K_*$, we can derive explicit bound for the state of the switched system. For $t \geq t_N$, the currently active controller is the RSP controller, $\hat{K}_t = K_N = K_*$. The behavior of the switched system (2.9) is then described by the constant matrices $\mathscr{A}_*$, $\mathscr{B}_* \doteq \begin{bmatrix} B_p & -B_pD_{k^*} \\ 0 & B_{k^*} \end{bmatrix}$ and $\mathscr{C}_* \doteq \begin{bmatrix} C_p & 0 \end{bmatrix}$ (assuming $D_p = 0$). The state transition matrix of the closed loop system (2.9) is $\Phi(t,t_{k^*}) = e^{\mathscr{A}_*(t-t_{k^*})}$. Owing to the exponential stabilizability of $\mathscr{P}$ by $K_*$, we have $||e^{\mathscr{A}_*(t-t_{k^*})}|| \leq ce^{-\lambda(t-t_{k^*})}$, for some positive constants $c, \lambda$. Applying the variation of constants formula to the state of the switched system $x$, we obtain:

$$\begin{aligned}
||x(t)|| &\leq ||e^{\mathscr{A}_*(t-t_{k^*})}||||x(t_{k^*})|| + \int_{t_{k^*}}^{t} ||e^{\mathscr{A}_*(t-\tau)}\mathscr{B}_*||||\omega||_\infty d\tau \\
&\leq ce^{-\lambda(t-t_{k^*})}||x(t_{k^*})|| + \mathscr{B}_*\frac{c}{\lambda}(1-e^{\lambda(t-t_{k^*})})||\omega||_\infty \\
&\leq ce^{-\lambda(t-t_{k^*})}||x(t_{k^*})|| + \mathscr{B}_*\frac{c}{\lambda}||\omega||_\infty
\end{aligned}$$

From the Equation C.2 in Appendix, $||\mathscr{C}_* x(t_{k^*})|| < \infty$, and so $||x(t_{k^*})|| < \infty$. Thus, $||x(t)||$ is bounded for all $t$, and

$$\lim_{t \to \infty} ||x(t)|| \leq \mathscr{B}_* \frac{c}{\lambda} ||\omega||_\infty .$$

which provides the upper bound on the switched system's state.

## 2.6 Treatment of the Time-Varying Plants

Adaptive switching scheme considered above provides guarantees of stability and convergence for a general nonlinear plant (with arbitrary but bounded noise and/or disturbances), whose unknown dynamics are time-invariant. In other words, the adaptive algorithm aims toward finding (given sufficient excitation) a robust controller - one that stabilizes the time-invariant plant. To have a 'truly' adaptive control system, one that is able to track changes in a plant whose parameters are either slowly time-varying or subject to infrequent large jumps (*e.g.*, component-failure-induced), one needs to deemphasize importance of the old data, which may be not demonstrative of the current plant behavior. For slow parameter variations, one usually endows a cost function (control selection law, in general) with data windowing, or fading memory. For instance, the cost functional to be minimized usually has an integral term:

$$J(\varphi,t) = ||\varphi(t)||^2 + \int_0^t e^{-\lambda(t-\tau)} \cdot ||\varphi(\tau)||^2 d\tau \qquad (2.10)$$

where $\lambda$ is a small non-negative number ('forgetting factor'), and $\varphi$ may be a vector of output data, identification error [77] *etc*. In such situations, convergence to a particular robustly stabilizing controller (given sufficient excitation) is neither achieved nor sought after. The property of the cost function that is lost by data windowing is its monotonicity in time. As a consequence, we do not have uniform shrinking of the candidate controller set anymore. Recall that at each time instant, the $\varepsilon$-cost minimization hysteresis algorithm falsifies a subset of the original candidate controller set whose current unfalsified cost level exceeds $V_{true}(K_{RSP})$; the resulting unfalsified controller sets form a nested, uniformly in time shrinking set, non-empty due to the feasibility assumption. Discarding time monotonicity, previously falsified controllers may be selected as optimal ones. Guarantees of convergence and finiteness of switches on the time interval $[0,\infty)$ are lost, but certain stability properties are preserved under modified conditions. The type of instability induced by infinitely fast switching can be avoided if an arbitrary, bounded away from zero, positive ratcheting step $\varepsilon$ is used.

For stability analysis, the definitions pertaining to the frozen-time analysis of stability of time-varying plants are useful, similarly as in [117].

**Definition 2.11.** The unknown plant $P$ whose parameters are frozen at their values at time $t^*$ is denoted $P^{t^*}$, and the closed loop switched system $\Sigma$ with the plant $P^{t^*}$ is denoted $\Sigma^{t^*}$. The set of all possible output signals $z = [u,y]$ reproducible by the switching system $\Sigma^{t^*}$ is denoted $\mathbf{Z}^{t^*}$.

**Definition 2.12.** The adaptive control problem is said to be *feasible in the frozen-time sense* if a candidate controller set $\mathbf{K}$ contains at least one controller such that the system $\Sigma^{t^*}(K,\mathscr{P})$ is stable. A controller $K \in \mathbf{K}$ is said to be a *feasible controller* for $\Sigma^{t^*}$ if the system $\Sigma^{t^*}(K,\mathscr{P})$ is stable.

**Assumption 2.2.** The adaptive control problem associated with the switched system $\Sigma$ is feasible.

**Definition 2.13.** Stability of a system $\Sigma^{t^*}: w \mapsto z$ is said to be *unfalsified* by data $(w,z)$ if there exist $\beta, \alpha \geq 0$ such that (2.1) holds; otherwise, it is said to be *falsified*.

**Definition 2.14.** The closed loop switched system $\Sigma^{t^*}$ is associated with the *true cost* $V_{true}^{\Sigma^{t^*}}: \mathbf{K} \to \mathbb{R}_+ \cup \{\infty\}$, defined as $V_{true}^{\Sigma^{t^*}}(K) = \sup_{z \in \mathbf{Z}^{t^*}, \tau \geq t^*} V(K,z,\tau)$.

**Definition 2.15.** For $\Sigma^{t^*}$, a *robust optimal controller* $K_{RSP}^{\Sigma^{t^*}}$ is a feasible controller that minimizes the true cost $V_{true}^{\Sigma^{t^*}}$.

Owing to the feasibility assumption, at least one such $K_{RSP}^{\Sigma^{t^*}}$ exists $\forall t^* \in \mathbf{T}$, and $V_{true}^{\Sigma^{t^*}}(K_{RSP}^{\Sigma^{t^*}}) < \infty$.

**Definition 2.16.** Let $r$ denote the input and $z_d = \Sigma^{t^*}(\hat{K}_t, \mathscr{P})r$ denote the resulting plant data collected with $\hat{K}_t$ as the current controller. Consider the adaptive control system $\Sigma^{t^*}(\hat{K}_t, \mathscr{P})$ of Figure 2.1 with input $r$ and output $z_d$. The pair $(V, \mathbf{K})$ is said to be *frozen-time cost detectable* if, without any assumption on the plant $P$ and for every $\hat{K}_t \in \mathbf{K}$ with finitely many switching times, the following statements are equivalent:

- $V(K_N, z_d, t)$ is bounded as $t$ increases to infinity.
- Stability of the system $\Sigma^{t^*}(\hat{K}_t, \mathscr{P})$ is unfalsified by the input-output pair $(r, z_d)$.

**Theorem 2.2.** *Consider the switched zero-input zero-output system $\Sigma: \mathfrak{L}_{2e} \to \mathfrak{L}_{2e}$ in Figure 2.1 where the unknown plant is time-varying. The input to the system is $w = [r\ d\ n\ x_0]^T$ (the fictitious input corresponding to K is $\tilde{w}_K = [\tilde{v}_K\ \tilde{d}_K\ \tilde{n}_K\ \tilde{x}_{0_K}]^T \in \tilde{W}_K(z)$), where $\tilde{v}_K(t) = N_K y(t) + D_K u(t)$ is the corresponding MFD), and the output is $z = [u\ y]^T$. Given is a cost function of the type*

$$V = \max_{\tilde{w}_K \in \tilde{W}_K(z)} \frac{\int_0^t e^{-\lambda_1(t-\tau)}||z(\tau)||^2 d\tau}{\int_0^t e^{-\lambda_2(t-\tau)}||\tilde{w}_K(\tau)||^2 d\tau} + \beta \tag{2.11}$$

*with $V = \beta$ when $t = 0$, where $\lambda_2 \geq \lambda_1 > 0$ are the weights on the past values of input and output data, respectively ('forgetting factors'), and $\beta$ is an arbitrary positive constant. Then, the cost function $V(K,z,t)$ is cost detectable in the frozen time sense, but not in general monotone in t.*

*Proof.* If stability of the system $\Sigma$ with $K$ in the loop is falsified, then for some $\tilde{w}_K \in \tilde{W}_K(z)$ there do not exist $\beta, \alpha \geq 0$ such that $||z||_t < \beta||\tilde{w}_K||_t + \alpha$, where

$||z||_t = ||(u,y)||_t \doteq \sqrt{||u||_t^2 + ||y||_t^2}$, and $||\zeta||_t \doteq \sqrt{\int_0^t e^{\lambda\tau}||\zeta(\tau)||^2 d\tau}$ is $e^{\lambda t}$-weighted, $\mathfrak{L}_2$-induced norm of a signal $\zeta(t)$ (in general, norm weight for the output need not coincide with the norm weight for the input signal). In particular, it follows that:

$$
\begin{aligned}
\infty = \limsup_{t\to\infty} \frac{||z||_t^2}{||\tilde{w}_K||_t^2} &= \limsup_{t\to\infty} \frac{\int_0^t e^{\lambda_1\tau}||z(\tau)||^2 d\tau}{\int_0^t e^{\lambda_2\tau}||\tilde{w}_K(\tau)||^2 d\tau} \\
&\le \limsup_{t\to\infty} \frac{\int_0^t e^{-\lambda_1(t-\tau)}||z(\tau)||^2 d\tau}{\int_0^t e^{-\lambda_2(t-\tau)}||\tilde{w}_K(\tau)||^2 d\tau} \\
< \limsup_{t\to\infty} \max_{\tilde{w}_K \in \tilde{W}_K(z)} &\frac{\int_0^\tau e^{-\lambda_1(t-\tau)}||z(\tau)||^2 d\tau}{\int_0^\tau e^{-\lambda_2(t-\tau)}||\tilde{w}_K(\tau)||^2 d\tau} + \beta + \gamma||K||^2 \\
&= \limsup_{t\to\infty} V(K,z,t) .
\end{aligned}
$$

From the last equation, cost detectability holds. □

In this case, since the unknown plant is varying in time, or the operating conditions are changing, the switching may never stop. However, the number of switches on any finite time interval can be derived. To this end, we apply the "pigeon-hole" lemma (similarly as in [46]), to find an upper bound on the number of switches on an arbitrary finite time interval. Recall from the Algorithm 2.1 that a currently active controller $\hat{K}$ will be switched out of the loop at time $t$ if its associated cost satisfies $V(\hat{K},z,t) \ge \varepsilon + \min_K V(K,z,t)$. We observe that this switching criterion will not be affected if we multiply both of its sides by some positive function of time $\Theta(t)$:

$$\Theta(t)V(\hat{K},z,t) \ge \varepsilon\Theta(t) + \Theta(t)\min_K V(K,z,t) .$$

Though we use the non-monotone in time cost function $V$ as in (2.11) in the actual switching algorithm, for analysis purposes we use its scaled version $V_m(K,z,t) \doteq \Theta(t)V(K,z,t)$, where the positive function of time $\Theta(t) \in \mathfrak{L}_{2e}$ is chosen so as to make $V_m(K,z,t)$ monotone increasing in time. The switching condition in the Algorithm 2.1 will also be modified to $V(\hat{K}_t,z,\tau) \ge \varepsilon + \min_{1\le i\le N} \min_{K\in B_\delta^i} V(K,z,\tau)$ where $B_\delta^i$ is the $i^{th}$ $\delta$-ball in the finite cover of $\mathbf{L}$, and $N$ is their number. This yields a hierarchical hysteresis switching similar to the one proposed in [46].

**Proposition 2.1.** *Consider the system* $\Sigma : \mathfrak{L}_{2e} \to \mathfrak{L}_{2e}$ *from the preceding sections, the cost (2.11), and a general candidate controller set* $\mathbf{K}$. *Suppose that the switching algorithm is chosen to be the hierarchical additive hysteresis switching. Then, the number of switches* $\aleph_{(t_0,t)}$ *on any finite time interval* $(t_0,t)$, $0 \le t_0 < t < \infty$ *is bounded above as:*

$$\aleph_{(t_0,t)} \le 1 + N + \frac{N}{\varepsilon\Theta(t)}(V_m(K,z,t) - \min_{K\in\mathbf{K}} V_m(K,z,t_0)) .$$

*Proof.* For brevity, we will omit $z_d$ from $V(K,z_d,t)$ in the sequel. Suppose that $K_{t_k}$ is switched in the loop at time $t_k$, and remains there until time $t_{k+1}$. Then,

$$V_m(K_{t_k},t) \leq \varepsilon\Theta(t) + V_m(K,t) \quad \forall t \in [t_k, t_{k+1}],\ \forall K \in \mathbf{K}\,.$$

Since $K_{t_k}$ was switched at time $t_k$, we also have $V_m(K_{t_k},t_k) \leq V_m(K,t_k)\ \forall K \in \mathbf{K}$. Owing to continuity of $V$ in time, $V_m(K_{t_k},t_{k+1}) = \varepsilon\Theta(t) + V_m(K_{t_{k+1}},t_{k+1})$. Now consider the non-trivial situation when more than one controller is switched in the loop on the time interval $(t_0,t)$. Then, some $K_q \in \mathbf{K}$ must be active in the loop at least $\nu \geq \frac{\aleph_{(t_0,t)}-1}{N}$ times ($N$ is the number of $\delta$-balls in the cover of $\mathbf{L}$). Denote the time intervals when $K_q$ is active in the loop as $[t_{k_1},t_{k_1+1}), [t_{k_2},t_{k_2+1}), ..., [t_{k_\nu},t_{k_\nu+1})$. According to the properties of the switching algorithm and monotonicity of $V_m$ we have:

$$\begin{aligned} V_m(K_q,t_{k_i+1}) &= \varepsilon\Theta(t) + V_m(K_{t_{k_i+1}},t_{k_i+1}),\quad i \in \{1,2,...,\nu-1\} \\ &\geq \varepsilon\Theta(t) + V_m(K_{t_{k_i+1}},t_{k_i}) \\ &\geq \varepsilon\Theta(t) + V_m(K_q,t_{k_i})\,. \end{aligned}$$

Also, because the switching time intervals are nonoverlapping, $V_m(K_q,t_{k_{i+1}}) \geq V_m(K_q,t_{k_i+1})$ and so $\varepsilon\Theta(t) + V_m(K_q,t_{k_i}) \leq V_m(K_q,t_{k_{i+1}})$. Since this holds $\forall i \in \{1,2,...,\nu-1\}$, we obtain, $\forall K \in \mathbf{K}$:

$$\begin{aligned} &(\nu-1)\varepsilon\Theta(t) + V_m(K_q,t_{k_1}) \leq V_m(K_q,t_{k_\nu}) \\ &\Rightarrow (\nu-1)\varepsilon\Theta(t) + V_m(K_q,t_0) \leq V_m(K,t)\,. \end{aligned}$$

Therefore, since $\aleph_{(t_0,t)} \leq \nu N + 1$, we derive, $\forall K \in \mathbf{K}$:

$$\aleph_{(t_0,t)} \leq 1 + N + \frac{N}{\varepsilon\Theta(t)}(V_m(K,z,t) - \min_{K\in\mathbf{K}} V_m(K,z,t_0))$$

which is finite since $\varepsilon > 0$, $\Theta(t)$ is a positive, $\mathfrak{L}_{2e}$ function of time, and $\min_{K\in\mathbf{K}} V_m(K,z,t_0))$ is finite due to feasibility assumption, as is $V_m(K,z,t)$ for some $K \in \mathbf{K}$. □

## 2.7 Behavioral Approach to Controller Unfalsification

A parallel between the controller unfalsification ideas on the one hand and the candidate elimination algorithm of Mitchell [70] on the other hand has been considered in [17], inspired by the results reported in [90]. It is noted that, as a special variant of Mitchell's elimination algorithm, controller falsification works by identifying hypothetical controllers that are consistent with past measurement data. The interest in this approach to understanding the essence of the unfalsification ideas that constitute the basis for the safe switching control stems from the particular simplicity and parsimony of the ensuing mathematical formulation, revealing some essential issues in feedback and learning from data. In addition, it is shown that system identification problems can be also tackled using the same truncated space unfalsified control concept framework.

In [17], it is argued that truncated space unfalsified control results from working in truncated signals spaces obtained from the application of an observation operator. Unfalsified control problems cast in this manner allows for simple falsification conditions which in turn require a reduced number of computations.

### *Canonical Representation of Unfalsified Control in Truncated Spaces*

The unfalsified control concept is a precise formulation of the controller validation problem in a hypothesis-testing framework [90]. Given the control goal (performance specification) and a set of hypotheses (candidate controllers), it evaluates them against experimental data. Experimental data are seen as a particular realization of the plant, which often includes actuators and sensors, and their disturbances and noises. For this reason, an extended plant is defined to be the plant with actuators and sensors. Therefore the experimental data are a realization of the extended plant. Another important aspect is that *a priori* information about the plant is used to define the hypothesis set. Models and any other prior knowledge about the plant are used to design the candidate controllers (hypotheses). Figure 2.5 depicts a general representation for an unfalsified control system. It is composed of elements that define relations between signals. The unfalsified control system has an internal structure with three blocks: the controller architecture, the learning processor and the controller. The controller architecture acts as the interface for the extended plant and the controller, and provides the performance signals needed by the learning processor, which in turn evaluates the candidate controllers and selects the best one to be used by the controller. The engine of the unfalsified control learning process is the evaluation of hypothesis against experimental data. In this regard, an alternative definition of an unfalsified controller can be stated as follows.

**Definition 2.17.** Given measurement information (data), a controller (hypothesis) is said to be falsified if the combined information (hypothesis and data) is sufficient to deduce that the performance specification (goals) would be violated if that controller were in the feedback loop. Otherwise, it is said to be unfalsified.

**Truncated Space Problem Formulation.** The general unfalsified control problem consists of evaluating if the candidate controllers are falsified by the currently available experimental data. The elements needed to define and solve the problem of evaluating if a control law is falsified are presented next. The basic components are signals and blocks. There are two groups of signals, *manifest* and *latent*, as in the behavioral approach of Willems [112]. Manifest signals ($\mathbf{Z}_{hypotheses} = \cup_i \mathbf{Z}_{hypothesis} \subset P_\tau \mathbf{Z}$) are the signals that manifest the observed behavior of the extended plant. These signals are control ($u$) and measurement ($y$) signals. Hence the manifest subspace is the input-output space, *i.e.* $U \times Y$ . Latent signals ($z_{latent}$) are signals internal to

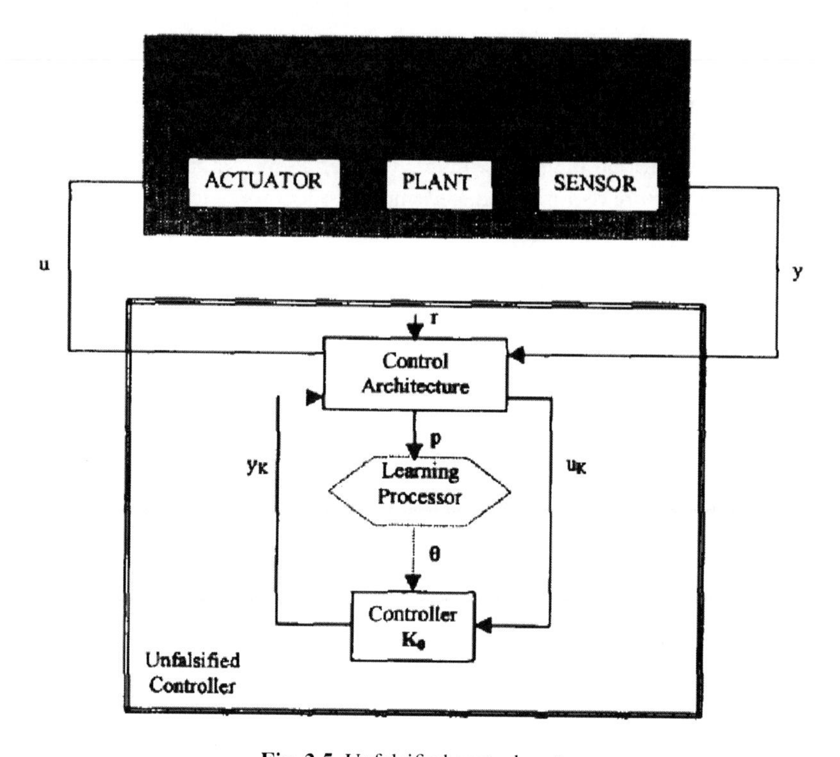


**Fig. 2.5** Unfalsified control system

the unfalsified controller used to define and evaluate performance and to produce the control signal. These are controller signals $(u_K, y_K)$, command signals $(r)$ and performance $(p)$ signals, and controller gains $(\theta)$. Consider the vector space formed from stacking all signals into $z = (z_{manifest}, z_{latent}) = (u, y, r, p, u_K, y_K, \theta)$. In addition, define the observation operator, $P_\tau$, which maps input-output signals to measurement signals. Examples of this operator are the time truncation operator or the sampling operator introduced in Section 2.2. Now define a vector space of truncated signals obtained from applying the observation operator to the vector space $\mathbf{Z}$.

**Definition 2.18.** Given a space $\mathbf{Z}$, and an observation operator $P_\tau$. The observations truncated signal space is defined as $Z^\tau = P_\tau \mathbf{Z}$.

The above definition provides a basis for working in the truncated signal space which is advantageous for practical reasons. Specifically, consider the following sets in the truncated signal space defined above:

- Goal set, $Z^\tau_{goal} \in P_\tau \mathbf{Z}$ denotes the performance specification.This could be described in terms of an observation-dependent error functional, $J(P_\tau \mathbf{Z})$, which evaluates the error between the actual system response and the desired one. Without loss of generality one can normalized it to $[0, 1]$. Hence,

$$Z^{\tau}_{goal}\{P_\tau \mathbf{Z} | 0 \leq J(P_\tau \mathbf{Z}) \geq 1\} \subset P_\tau \mathbf{Z} \tag{2.12}$$

Since the functional evaluates the error, the value of the cost is a measure of the performance such that smaller cost indicates better performance. This cost can be used to define a partial ordering of hypotheses.

- Controller hypothesis, $Z^{\tau}_{hypothesis}(\theta) \in P_\tau \mathbf{Z}$, denotes the graph of a $\theta$-dependent dynamical control law $K_\theta(P_\tau \mathbf{Z}) = 0$, *viz.*,

$$Z^{\tau}_{hypothesis}(\theta) = \{P_\tau z | K_\theta(P_\tau z) = 0\} \subset P_\tau \mathbf{Z} \tag{2.13}$$

- Hypotheses set,

$$Z^{\tau}_{hypothesis}(\theta) : \Theta \to 2^{P_\tau \mathbf{Z}} \tag{2.14}$$

denotes the set of candidate controllers parameterized by a vector $\theta$.

- Data set, $Z^{\tau}_{data} \in P_\tau \mathbf{Z}$, is defined as

$$Z^{\tau}_{data} = \{P_\tau z | P_\tau (z_{manifest}) = (u,y)_{measurements}\} \subset P_\tau \mathbf{Z} \tag{2.15}$$

where $(u,y)_{measurements}$ represents past plant input-output observed data available at time $\tau$.

- The unfalsification goal:

$$(Z^{\tau}_{data} \cap Z^{\tau}_{hypothesis}(\theta) \subset Z^{\tau}_{goal}) \tag{2.16}$$

These truncated sets are used in [17] to test falsification for purposes of controller validation, adaptive control and system identification.

**Problem 1: Truncated Space Unfalsified Control Problem**

Given a performance specification (goal set, $\mathbf{Z}^{\tau}_{goal} \neq \varnothing$), a set of candidate controllers (hypotheses, $\mathbf{Z}_{hypotheses} = \cup_i \mathbf{Z}_{hypothesis}(\theta) \neq \varnothing$), and experimental data ($\mathbf{Z}^{\tau}_{data} \neq \varnothing$), determine the subset of candidate controllers (hypotheses) which is not falsified. This formulation brings out important issues, some of which are that every element of the truncated space data set is not only consistent with the observed data $(u,y)_{measurements} \in P_{\mathbf{Z}_{manifest}} \mathbf{Z}^{\tau}_{data}$ (where $P_{\mathbf{Z}_{manifest}}$ denotes the projection $z_{manifest} = P_{\mathbf{Z}_{manifest}} z$), but it is also the unique element of the set $P_{\mathbf{Z}_{manifest}} \mathbf{Z}^{\tau}_{data}$. This means that in testing unfalsification there is no need to analyze multiplicities of unseen future or intersample behaviors for the manifest signals $(u,y)$. Another advantage, as argued in [17], is that one has more flexibility in the design since instead of working in $\mathbf{R} \times \mathbf{Y} \times \mathbf{U}$, one deals with $\mathbf{Z}$, which may in general be a bigger signal space. This flexibility is very useful in the definition of the performance specification. For example it allows for specifications involving signals derived from a hypothesis-dependent reference model, which is not possible when performance goals must depend only on $(r,y,u)$.

**Results.** To build up the background for a solution to the truncated space unfalsified control problem, the concept of data-hypothesis consistency is defined, which considers whether a hypothesis (controller) connected to the plant in closed-loop could have produced the measurement data.

**Definition 2.19. Data-hypothesis consistency.** Given a truncated space unfalsified control problem, it is said that a hypothesis is consistent with the data if

$$P_{\mathbf{Z}_{manifest}} \mathbf{Z}^{\tau}_{data} \subset P_{\mathbf{Z}_{manifest}} \mathbf{Z}^{\tau}_{hypothesis}(\theta) . \tag{2.17}$$

Note that if a particular controller is not consistent with data, then irrespective of the goal the data cannot falsify this controller.

**Theorem 2.3.** *For a given truncated space unfalsified control problem, a candidate control law (hypothesis, $\mathbf{Z}^{\tau}_{hypothesis}(\theta)$) consistent with the experimental data is unfalsified by data $P_{\mathbf{Z}_{manifest}} \mathbf{Z}^{\tau}_{data}$ if and only if $\mathbf{Z}^{\tau}_{data} \cap \mathbf{Z}^{\tau}_{hypothesis}(\theta) \cap \bar{\mathbf{Z}}^{\tau}_{goal} \neq \varnothing$, where $\bar{\mathbf{Z}}^{\tau}_{goal}$ is the complement of $\mathbf{Z}^{\tau}_{goal}$. Otherwise, it is falsified.*

Then, a solution to Problem 1 is achieved by testing each candidate controller for unfalsification via Theorem 2.3. Applied to an adaptive control problem, the task formulation can be stated as follows.

## *Adaptive Truncated Space Unfalsified Control*

Given a $\gamma$-dependent goal set $\mathbf{Z}^{\tau}_{goal}(\gamma) \neq \varnothing$, $\gamma \in \mathbb{R}$, a set of candidate controller hypotheses, $\mathbf{Z}^{\tau}_{hypotheses} = \cup \mathbf{Z}^{\tau}_{hypothesis}(\theta) \neq \varnothing$, and an evolving $\tau$-dependent experimental data set ($\mathbf{Z}^{\tau}_{data} \neq \varnothing$), then at each time $\tau$ find the least $\gamma = \gamma_{opt}$ for which the set of unfalsified controllers is non-empty and select a controller $K_{\gamma_{opt}}$ from this set.

Figure 2.6 represents a canonical representation of the above defined adaptive control problem. The role of the learning processor is twofold: updating the unfalsified candidate controller subset and selecting the best controller to be put into the loop. The role of the parameter $\gamma$ in the goal set is to define a partial ordering of the hypotheses.

$$\mathbf{Z}^{\tau}_{goal}(\gamma) = \{P_{\tau} z | 0 \leq J(P_{\tau} z) \leq \gamma\} . \tag{2.18}$$

The adaptive unfalsified control problem viewed from the optimization aspect can be stated as follows:

**Theorem 2.4.** *The solution to an adaptive truncated space unfalsified control problem is given by the following constraint optimization: At each time $\tau$, find a controller $K_{\theta_{opt}}(\mathbf{Z}_{hypothesis}(\theta_{opt}))$, which solves:*

$$\gamma_{opt} \triangleq \arg\min_{\mathbf{Z}^{\tau}_{hypothesis}(\theta)} (\arg\min_{P_{\tau} z} J(P_{\tau} z)) \tag{2.19}$$

*such that*

$$\begin{aligned} P_{\tau} z \in \mathbf{Z}^{\tau}_{data} \cap \mathbf{Z}^{\tau}_{hypothesis}(\theta), \\ \mathbf{Z}^{\tau}_{data} \cap \mathbf{Z}^{\tau}_{hypothesis}(\theta) \subseteq \mathbf{Z}^{\tau}_{goal}(\gamma) \;\; \textit{and} \\ \mathbf{Z}^{\tau}_{hypothesis}(\theta) \in \mathbf{Z}^{\tau}_{hypothesis} . \end{aligned} \tag{2.20}$$

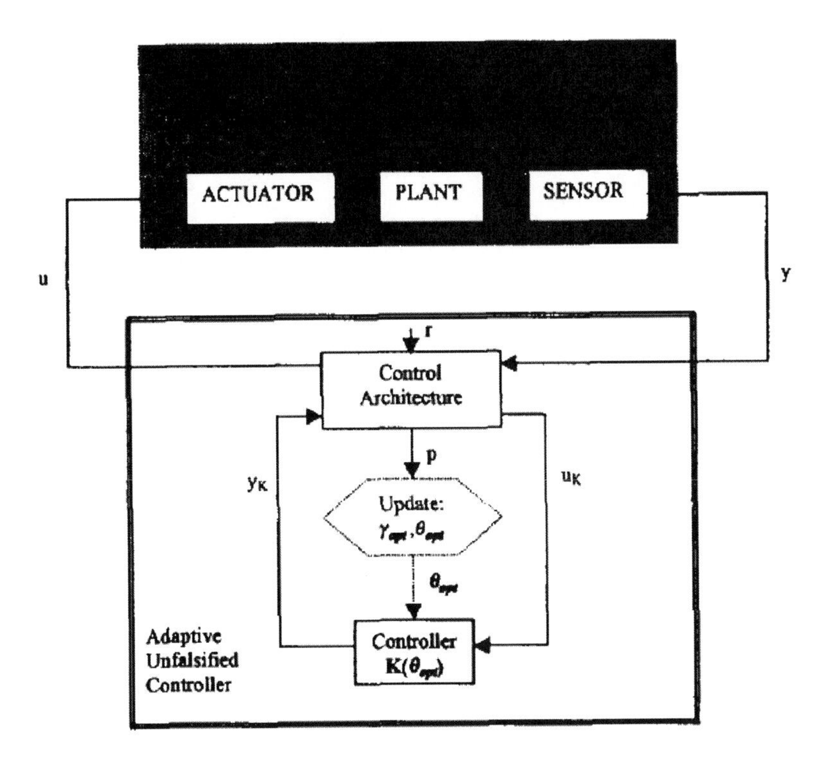


**Fig. 2.6** Adaptive Unfalsified Control System

With respect to the practical considerations of this canonical representation of the unfalsified adaptive control in time truncated space, an important remark is that the unfalsified control could be run just once, or it can be run iteratively on evolving past data as each new datum is acquired. If run iteratively on evolving past data, then it can be used for real time controller adaptation. Adaptive updates of the current controller can be done either periodically or aperiodically, or by a mixture of both. If run periodically, then the algorithm rate can be arbitrary. For quickly varying systems the highest possible rate (sensor sampling rate) may be preferred, but for slowly varying systems one would settle for a lower rate. For an aperiodical run, a performance-dependent function should be defined to decide when to run it.

Figure 2.7 shows an algorithm for a generic unfalsified learning processor. Every time the learning processor is called, the data available defines the observations operator which in turn helps to define the truncated signal space (Definition 2.17) in which to work. The "Choose the best" block identifies the best hypothesis to be used in the adaptive (and also identification) problems.

The basic components of this algorithm are defined in the truncated signal space obtained from the use of the observations operator.

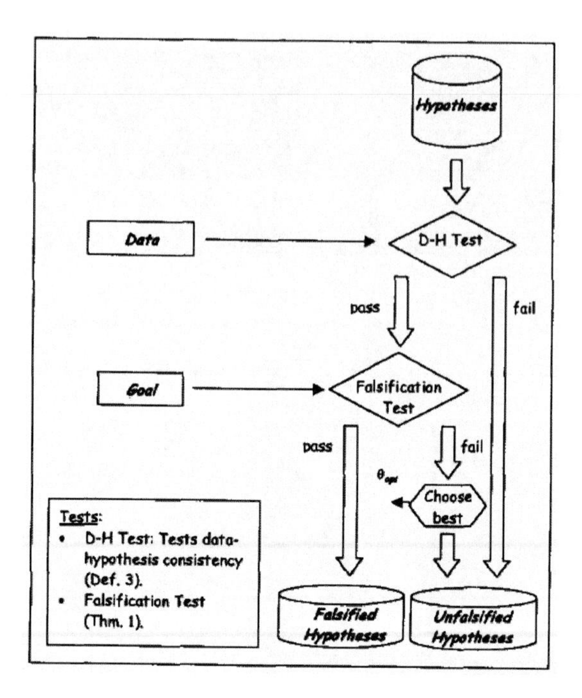


**Fig. 2.7** Unfalsified learning processor

- *Goal*: A cost function measuring the error between the actual and desired performances represents the goal. As an example, it may have the form of an inequality. A weighted $l_{2e}$ norm specification of the mixed sensitivity type is a common example. For real-time application with a time varying system, a forgetting factor is often used to rest importance to the old measurements.
- *Hypotheses*: Recent research has alleviated the previous restriction of the controllers' (or models') hypotheses to the causally left invertible form. With this hindrance lifted, there are various forms of the hypotheses that can be used; the unfalsification-based system identification, for example [58], uses ARX parametrization.
- *Data*: The experimental data available defines the observations operator. The data set is the projection of experimental data onto the latent variable subspace of the truncated signal space. Typically, experimental data will be discrete signals of finite length.

# Chapter 3
# Safe Switching Adaptive Control: Examples

**Abstract.** In this chapter, we present examples that illustrate the stability robustness properties of the safe adaptive control algorithm in the high uncertainty setting, and compare it with alternative control schemes. The motivating example gives insight into the model mismatch stability failure associated with model based adaptive control schemes lacking the necessary properties of their cost function, and a solution to the preceding problem is provided. Following this example, transient behavior improvement of the safe adaptive schemes is discussed. For completeness, some applications of the unfalsified control concept are reproduced in the last section of this chapter.

## 3.1 Benchmark Example

The Algorithm 2.1 in Section 2.5 originated as the hysteresis switching algorithm in [75]. We emphasized that the power of the hysteresis switching lemma was clouded in the cited study by imposing unnecessary assumptions on the plant in the demonstrations of the algorithm functionality. One of the plant properties required in [75] for ensuring 'tunability' was the minimum phase property of the plant. We have shown in theory that the cost detectability property is assured by properly choosing a cost function, and is not dependent on the plant or exogenous signals. In the following, we present a MATLAB® simulation example that demonstrates these findings is presented.

Assume that a true, unknown plant transfer function is given by $\mathscr{P}(s) = \frac{s-1}{s(s+1)}$. It is desired that the output of the plant behaves as the output of the stable, minimum phase reference model $\mathscr{P}_{ref}(s) = \frac{1}{s+1}$. Given is the set of three candidate controllers: $C_1(s) = -\frac{s+1}{s+2.6}$, $C_2(s) = \frac{-s+1}{0.3s+1}$ and $C_3(s) = -\frac{s+1}{-s+2.6}$, each of which stabilizes a different possible plant model. The task of the adaptive control is to select one of these controllers, based on the observed data. The problem is complicated by the fact that in this case, as is often the case in practice, the true plant $\mathscr{P}(s)$ is not in the model set, *i.e.*, there exists a 'model mismatch'.

M. Stefanovic and M.G. Safonov: Safe Adaptive Control, LNCIS 405, pp. 41–128.
springerlink.com 

A simple analysis of the non-switched system (true plant in feedback with each of the controllers separately) shows that $C_1$ is stabilizing (yielding a non-minimum phase but stable closed loop) while $C_2$ and $C_3$ are destabilizing. Next, a simulation was performed of a switched system (see Figure 3.1), where Algorithm 2.1 was used to select the optimal controller, and the cost function was chosen to be a combination of the instantaneous error and a weighted accumulated error:

$$J_j(t) = \tilde{e}_j^2(t) + \int_0^t e^{-\lambda(t-\tau)}\tilde{e}_j^2(\tau)d\tau, \quad j = 1,2,3\,. \tag{3.1}$$

In the preceding, $\tilde{e}_j$ is the fictitious error of the $j^{th}$ controller, defined as

$$\tilde{e}_j = \tilde{y}_j - y \tag{3.2}$$

where $\tilde{y}_j = \mathscr{P}_{ref}\tilde{r}_j$ and $\tilde{r}_j = y + C_j^{-1}u$, and where $\mathscr{P}_{ref} = \frac{1}{s+1}$ is the stable, minimum phase reference model. The plant input and output signals, $u$ and $y$, respectively, constitute the closed loop system output signals. The cost function above is the same cost function used in the multiple model switching adaptive control scheme [77], in which $\tilde{e}_j$ was replaced by $e_{I_j}$, the identification error of the $j^{th}$ plant model (for the special case of the candidate controllers designed based on the model reference adaptive control (MRAC) method, $e_{I_j}$ is equivalent to the control error and to the fictitious error (3.2)).

The simulations assume a band-limited white noise at the plant output and the unit-magnitude square reference signal. The stabilizing controller $C_1$ has initially been placed in the loop, and the switching, which would normally occur as soon as the logic condition of Algorithm 2.1 is met, is suppressed during the initial 5 seconds of the simulation. That is, the adaptive control loop is not closed until time $t = 5$ $s$. The reason for waiting some period of time before engaging the switching algorithm will be explained shortly. The forgetting factor $\lambda$ is chosen to be 0.05. Figures 3.2 and 3.4 show the cost dynamics and the reference and plant outputs, respectively.

Soon after the switching was allowed (after $t = 5$ $s$), the algorithm using cost function (3.1) discarded the stabilizing controller initially placed into the loop and latched onto a destabilizing one, despite the evidence of instability found in data. Even though the stabilizing controller was initially placed in the loop, and forced to stay there for some time (5 seconds in this case), as soon as the switching according to (3.1) was allowed, it promptly switched to a destabilizing one. This model-match instability happens because the cost function (3.1) is not cost detectable. Note that the initial idle period of 5 seconds is used only to emphasize that, even when the data are accumulated with the *stabilizing* controller, the switching algorithm based on (3.1) can disregard these data, and latch onto a destabilizing controller. This idle period is *not the dwell time* in the same sense used in the dwell-time switching control literature, *e.g.*, [26], [42], [43].

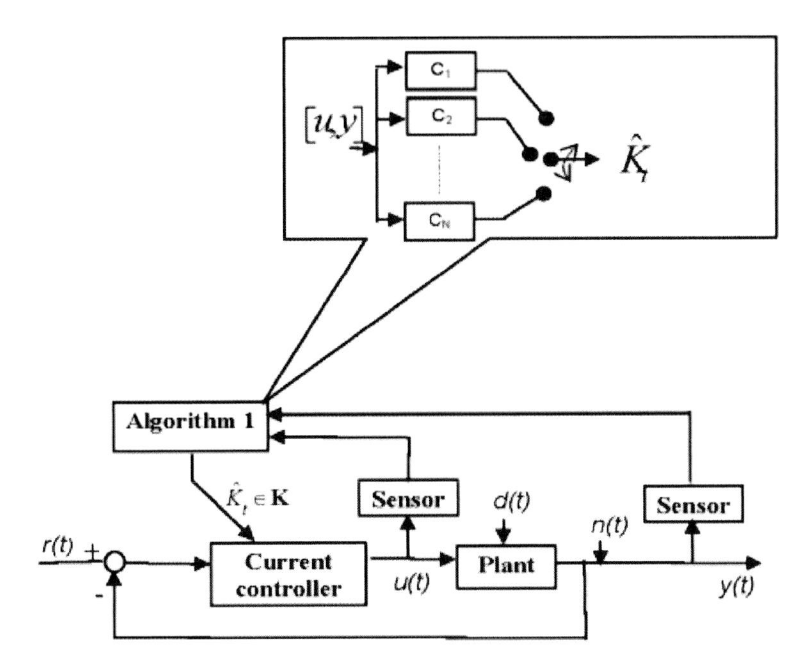


**Fig. 3.1** Switching feedback control system configuration

Next, a simulation was performed of the same system, but this time using a cost detectable cost function (*i.e.*, one that satisfies the conditions of Theorem 1):

$$V(K,z,t) = \max_{\tau \in [0,t]} \frac{||u||_\tau^2 + ||\tilde{e}_K||_\tau^2}{||\tilde{v}_K||_\tau^2 + \alpha} \tag{3.3}$$

*viz*, an $\mathfrak{L}_{2e}$-gain type cost function (factor $\gamma||K||^2$ needed for coerciveness is not necessary, since the set of candidate controllers is finite in this example). Cost detectability of $(V, \mathbf{K})$ follows from Lemma 2.5. The modified fictitious reference signal $\tilde{v}_K$ is used, instead of $\tilde{r}_K$, because of the presence of the non-minimum phase controller $C_2$. It is calculated from the on-line data as $\tilde{v}_K = D_K u + N_K y$ (according to (B.1)), where $D_{K_1} = \frac{s+2.6}{s+2}$ and $N_{K_1} = -\frac{s+1}{s+2}$ for $C_1$; $D_{K_2} = \frac{0.3s+1}{s+2}$ and $N_{K_2} = \frac{-s+1}{s+2}$ for $C_2$; and $D_{K_3} = \frac{s-2.6}{s+2}$ and $N_{K_3} = \frac{s+1}{s+2}$ for $C_3$. The corresponding simulation results are shown in Figure 3.3 and Figure 3.5. The initial controller was chosen to be $C_3$ (a destabilizing one). The constant $\alpha$ was chosen to be 0.01.

The destabilizing controllers are kept out of the loop because of the properties of the cost function (cost detectability). For further comparison, the same simulation is repeated, with the destabilizing initial controller $C_3$, but the switching is allowed after 5 seconds (as in the first simulation run shown in Figure 3.2). It can be seen from Figure 3.6, that, despite the forced initial latching to a destabilizing controller, and therefore increased deviation of the output signal from its reference value, the algorithm quickly switches to a stabilizing controller, eventually driving the controlled output to its preferred value.

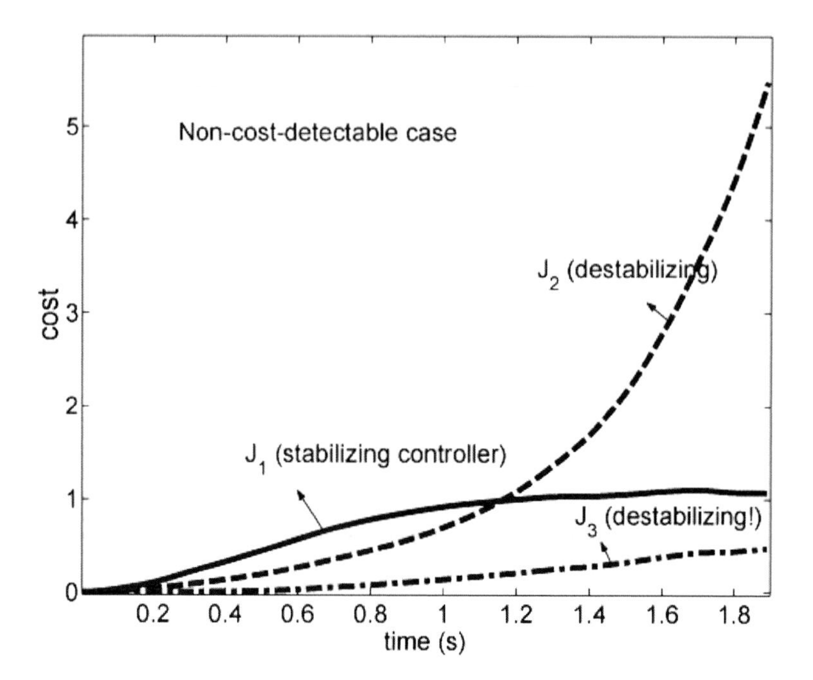


**Fig. 3.2** Cost function (3.1) shows the destabilizing property of the non cost detectable optimal cost-minimizing controller

The foregoing example shows that closing an adaptive loop that is designed using a non cost detectable cost function like (3.1) can destabilize, even when the initial controller is stabilizing. In the example, this happens because there is a large mismatch between the true plant and the plant models used to design the candidate controllers. On the other hand, Theorem 1 implies, and the example confirms, that such model mismatch instability cannot occur when the adaptive control loop is designed using the cost detectable $\mathfrak{L}_{2e}$-gain type cost function (3.3).

## 3.2 Further Comparison with the Multiple Model Based Switching Control

The case study that served as a source of inspiration for counterexamples in this monograph is found in the study of Narendra and Balakrishnan [9], [77]. In Chapter 3 of [9], a comparison is drawn between the indirect MRAC adaptive control using switching among multiple models, and the indirect adaptive control using hysteresis switching algorithm of Morse [72], [75]. It is stated that the choice of the performance index (cost function) used throughout [9], namely:

$$J_j(t) = \alpha e_j^2(t) + \beta \int_0^t exp(-\lambda(t-\tau))e_j^2(\tau)d\tau, \; j = 1,2,3 \qquad (3.4)$$

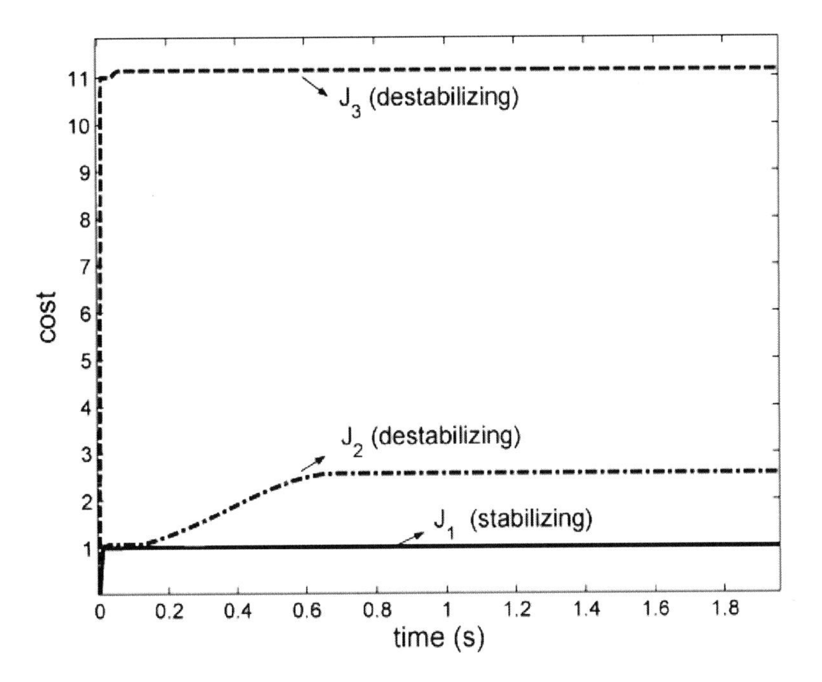


**Fig. 3.3** Cost function (3.3) for the cost detectable case

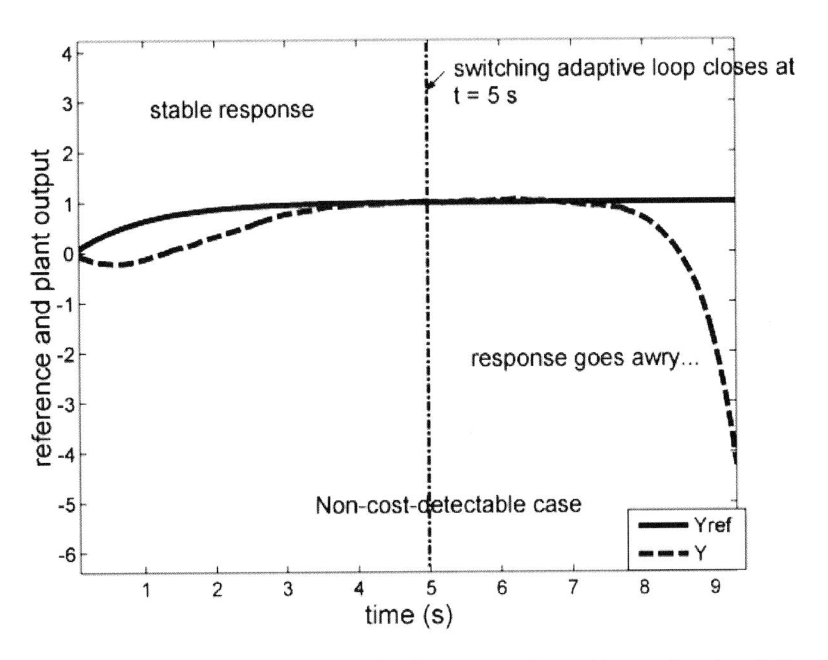


**Fig. 3.4** Reference and plant outputs for the non cost detectable cost function (3.1)

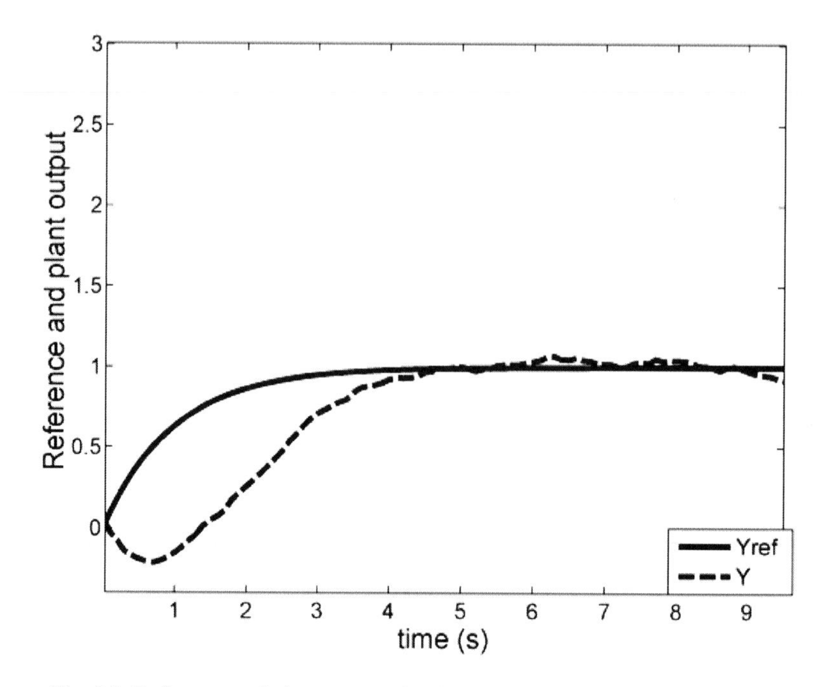


**Fig. 3.5** Reference and plant outputs for the cost detectable cost function (3.3)

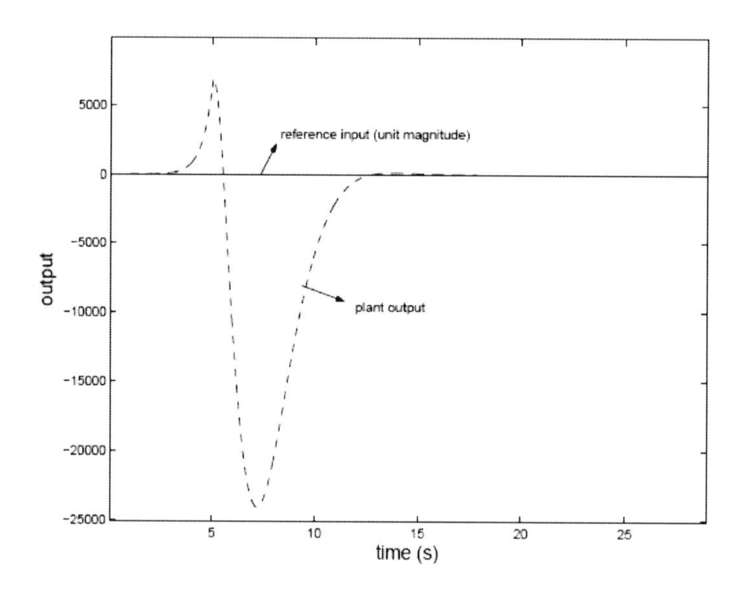


**Fig. 3.6** Reference and plant outputs using cost function (3.3) and the switching Algorithm 2.1, with the initial delay of 5 seconds

which incorporates both instantaneous and long term measures of accuracy, cannot be used in exactly the same form in the stability proof of the hysteresis switching lemma of [75], since the identification errors $e_j$ cannot be considered square integrable or even bounded, implying that the cost (3.4) does not satisfy the boundedness condition required for a stabilizing model. In [9], a modification to the above cost is proposed in the form of:

$$\bar{J}_j(t) = \int_0^t \frac{e_j^2(\tau)}{1+\bar{\omega}^T(\tau)\bar{\omega}(\tau)} d\tau, \quad j = 1,2,3 \tag{3.5}$$

where $\bar{\omega}$ is the on-line estimate of the vector $\bar{\omega}^*$ consisting of the plant input, output, and sensitivity vectors, which is familiar from the traditional MRAC theory for a linear time-invariant single-input single-output (LTI SISO) plant [76]. Such a modified cost function satisfies the stability and finite time switching conditions of [75]. The problem pointed out, however, was that this choice of the cost function is based strictly on the stability considerations, not performance, whereas the design parameters $\alpha, \beta$, and $\lambda$ in (3.4) are claimed to provide flexibility in optimizing performance.

In [9], the superiority of the cost index choice (3.4) over (3.5) is advocated and a simulation is furnished showing a substantially better transient response using the cost function (3.4) rather than (3.5). Though it may seem that the former cost index results in a superior performance, an important question was not considered: Are the conditions under which stability of the proposed control design is assured verifiable? We answer this question in the negative, since it is clear that one does not know *a priori* whether there exists a mismatch between the true plant and the proposed models. To provide better understanding, let us reconsider the problem presented in [9], Chapter 3, Section 3.3.3.

The LTI SISO plant to be controlled is assumed to have the transfer function (unknown to us) $G(s) = \frac{0.5}{s^2+0.05s+1}$. The control objective is to track the output of a reference model $W_m(s) = \frac{1}{s^2+1.4s+1}$ to a square wave reference input with a unit amplitude and a period of 4 units of time. Two (fixed) candidate models are considered: $W_{F1}(s) = \frac{1}{s^2+s-1.5}$ and $W_{F2}(s) = \frac{2}{s^2+2s+1}$. The model reference controllers are designed (off-line) for each of the two models, with the following parameter vectors: $\theta_{F1}^* = [1, \ -0.4, \ -2.5, \text{ and } -0.6]$ and $\theta_{F2}^* = [0.5, \ 0.6, \ -0.3, \text{ and } 0]$, where

$$u_{Fi}^* = \theta_{Fi}^* \cdot \underline{\omega} = [k^* \ \ \theta_1^* \ \ \theta_0^* \ \ \theta_2^*] \cdot \begin{bmatrix} r \\ \omega_1 \\ y \\ \omega_2 \end{bmatrix} \tag{3.6}$$

is the perfect match control law for the model $W_{Fi}$ resulting in perfect following of the reference command. The sensitivity vectors $\omega_1, \omega_2 : \mathbb{R}_+ \rightarrow \mathbb{R}^{2n}$, ($n$ order of the unknown plant) are defined, as in the standard MRAC problem, as:

$$\begin{aligned} \dot{\omega}_1 &= \Lambda \omega_1 + lu, \\ \dot{\omega}_2 &= \Lambda \omega_2 + ly \end{aligned} \tag{3.7}$$

with $(\Lambda, l)$ an asymptotically stable, controllable pair.

The resulting control actions for the two controllers are then calculated as:

$$\begin{aligned} u_{K_1} &= u^*_{F1} \frac{s+1}{s+1.4} r - \frac{2.5s+3.1}{s+1.4} y, \\ u_{K_2} &= u^*_{F2} \frac{s+1}{2(s+0.4)} r - \frac{0.3(s+1)}{s+0.4} y\,. \end{aligned} \tag{3.8}$$

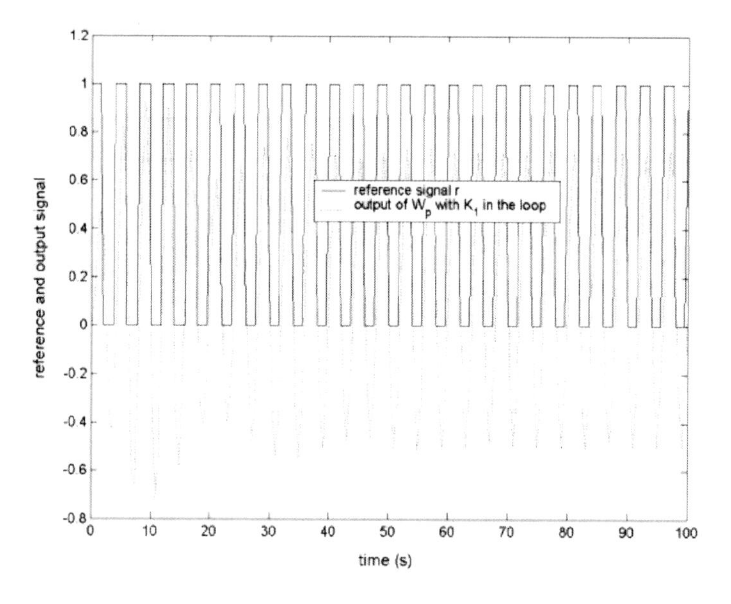


**Fig. 3.7** True plant in feedback with controller $K_1$

A non-switched analysis is then performed, with the real plant in feedback with each of the designed controllers separately, showing that $K_1$ is stabilizing, whereas $K_2$ is not (the closed loop has a pair of the right half plane (RHP) poles close to the imaginary axis); see Figures 3.7 and 3.8. In the next simulation, switching between these two controllers is performed using the performance index advocated in [9], [77]:

$$J_j(t) = e_j^2(t) + \int_0^t e_j^2(\tau) d\tau\,. \tag{3.9}$$

The results of the simulation are shown in Figure 3.9 (cost) and Figure 3.10 (output). The switching scheme of [77] using the criterion (3.9) gives preference to the destabilizing controller $K_2$ since the parameters of its corresponding plant model are closer to those of the real plant $G$. In [77], those authors attempt to avoid this pitfall by assuring sufficient density of the plant models, to increase the probability that the real plant falls within the robustness bounds of at least one candidate plant model, so that the corresponding robustly stabilizing controller of that model will also be robustly stabilizing for the real plant. While this idea is intuitively appealing, we should always insure that we do not deviate from a stabilizing and sufficiently well performing controller (when there exists one) and latch onto an even destabilizing one, as the example is this section demonstrates.

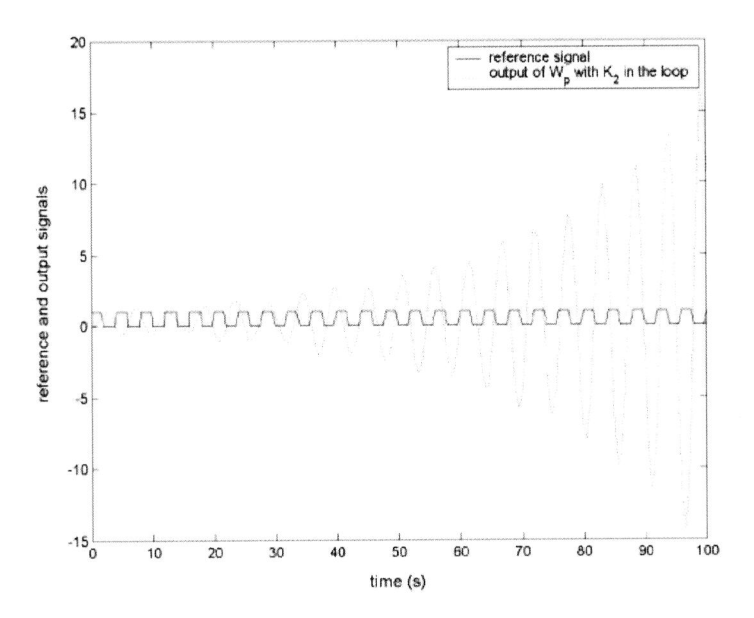


**Fig. 3.8** True plant in feedback with controller $K_2$

Now, let us consider the switching Algorithm 2.1 with the cost function from Morse *et al.* [75], which satisfies the conditions of the hysteresis switching lemma (repeated for expedited reading in (3.10)):

$$\bar{J}_j(t) = \int_0^t \frac{e_j^2(\tau)}{1+\bar{\omega}^T(\tau)\bar{\omega}(\tau)} d\tau, \quad j = 1,2,3\,. \tag{3.10}$$

Indeed, this type of a cost functional satisfies some of the properties required by the Theorem 1, namely it reaches a limit (possibly infinite) as $t \longrightarrow \infty$; at least one $\bar{J}_j(t)$ (in this case $\bar{J}_1(t)$) is bounded, and the indirect controller $C_1$ assures stability for the real plant. A hysteresis algorithm is employed with a hysteresis constant $\varepsilon > 0$ to

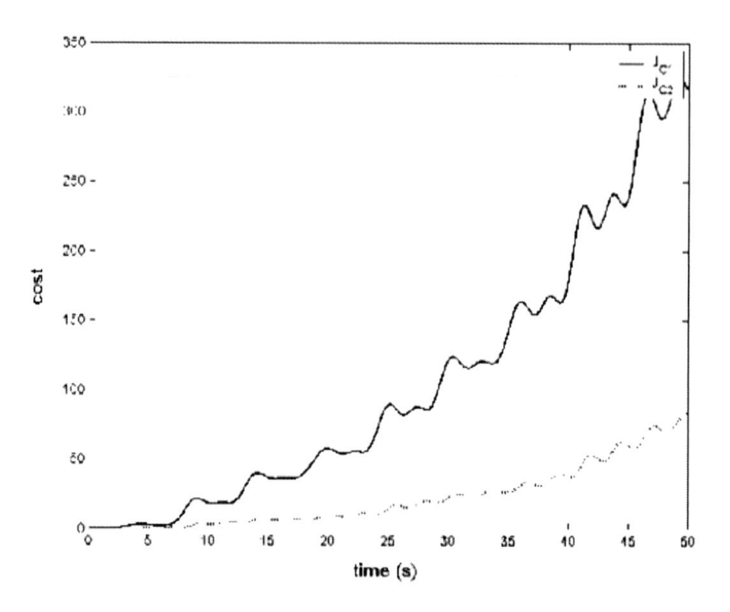


**Fig. 3.9** Cost function trajectory for controllers $K_1$ and $K_2$ (cost function (3.9))

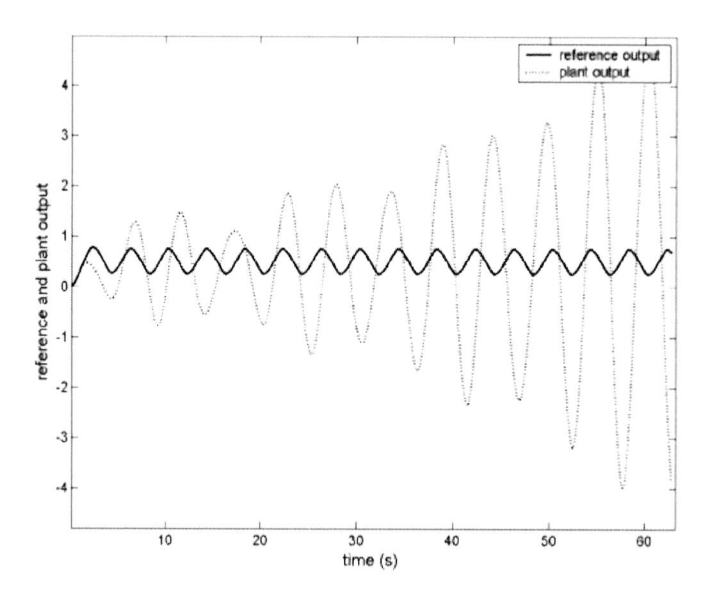


**Fig. 3.10** Reference and plant output using cost function (3.9)

prevent chattering. This should assure that the switching stops at the model $I_1$ for which $\bar{J_1}(t)$ is bounded, and so in turn assure stability. Let us look at the simulation results. Figure 3.11 shows the output of the plant, together with the preferred, reference model output. Apparently, the selector (cost) function of [75] still opted for the destabilizing controller. The explanation is ascribed to the lack of the cost detectability property of (3.10) - it employs a ratio of the error signal (expressed in terms of the system output signal) and a signal composed of the the plant input and output (in measured and filtered forms) - which are considered output signals from the standpoint of the overall closed loop switched system. Therefore, this form of the cost function does not detect instability. Costs of both controllers can be seen to be growing (Figure (3.12)), but $J_2$ is growing more slowly and is bounded above by $J_1$. Finally, consider the switching using the cost function:

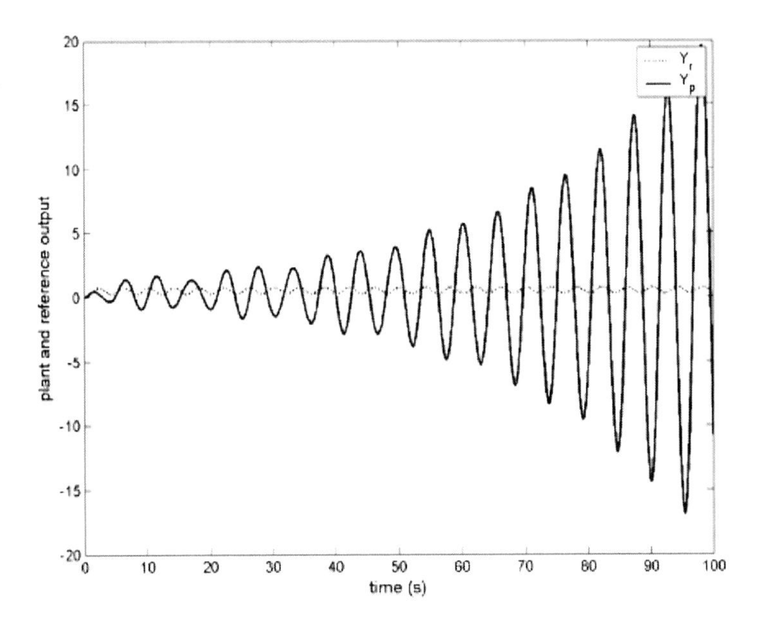


**Fig. 3.11** Reference and plant output using cost function (3.5)

$$V(K,z,t) = \max_{\tau \leq t} \frac{\tilde{e}_j^2(\tau) + \int_0^\tau \tilde{e}_j^2(\sigma)d\sigma}{\int_0^\tau \tilde{r}_j^2(\sigma)d\sigma} \tag{3.11}$$

which conforms to the conditions of the Theorem 1. The simulation results are shown in Figures 3.13 and 3.14. Although $C_2$ is initially switched in the loop, the switching algorithm recognizes its destabilizing property and replaces it with $C_1$.

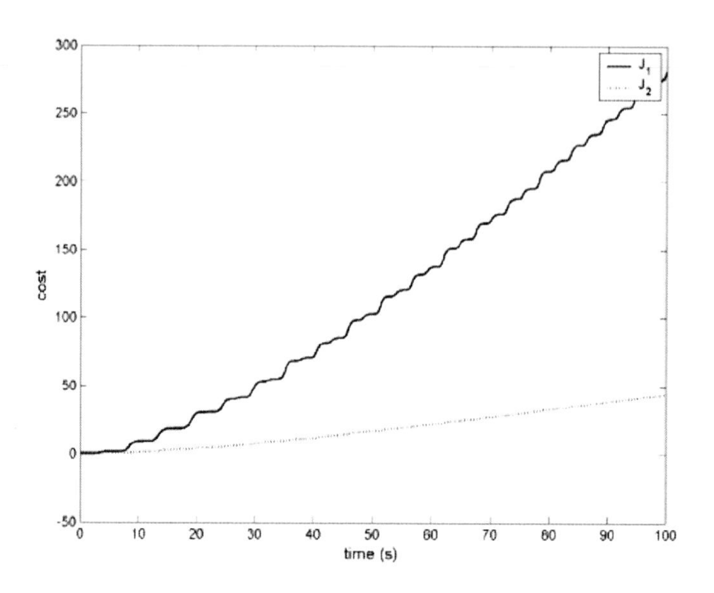


Fig. 3.12 Cost function for controller $C_1$ and $C_2$ using cost function (3.5)

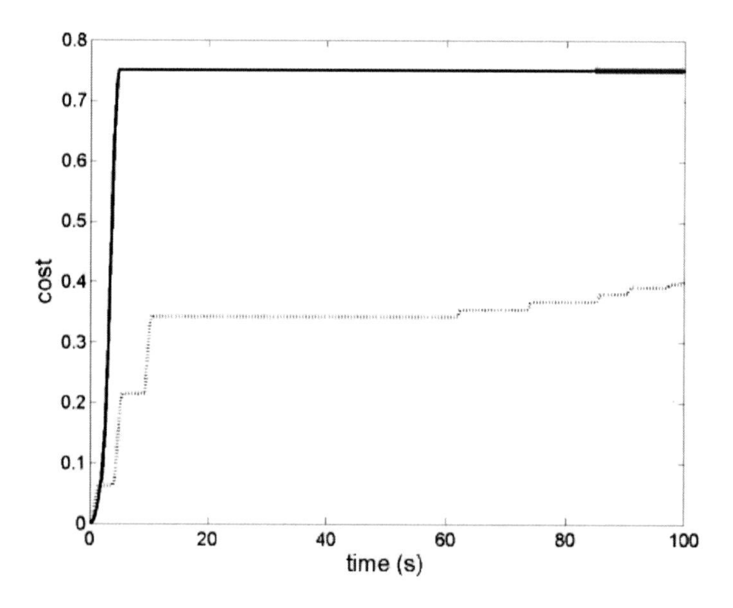


Fig. 3.13 Cost function for controller $C_1$ (dotted line) and $C_2$ (solid line), using cost function (3.9)

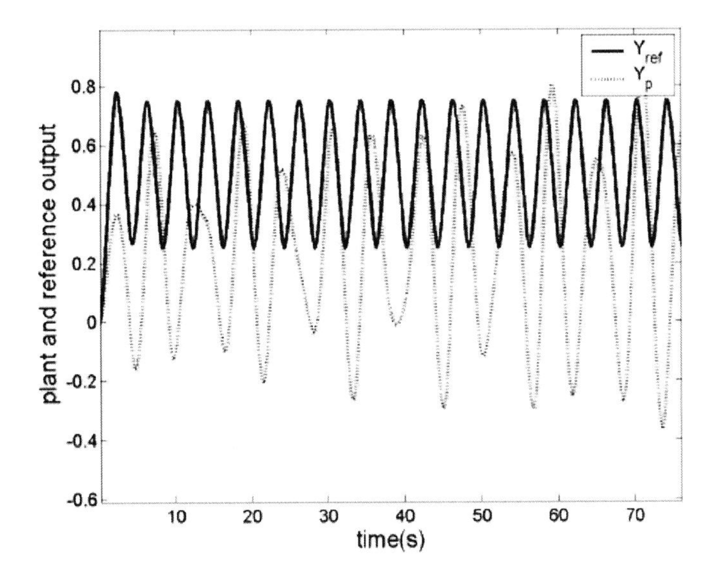


**Fig. 3.14** Reference and plant output using cost function (3.9)

## 3.3 Performance Improvement

The theory of the safe switching adaptive control design excludes the possibility of instability, *i.e.*, growing of the signals without a bound in the closed loop system, as long as a stabilizing controller exists in the candidate set. In this manner, it overcomes the limitation of the model based stability results that have plagued many of the existing adaptive control schemes. A notable remark has been raised in [26] concerning the quality of the system's response in between the switching instants, *i.e.*, closed loop system's performance. In [22], bandpass filters were proposed to filter the signals involved in the cost function, inspired by the results in [87], so as to improve the ability of the cost function to more promptly recognize the instability trend caused by the current controller in the loop. The proposed modification in the cost function, or rather in the signals constituting the cost function, yields better performance than the dwell time imposed in between switches in [26]. The concern about the dwell time that the closed loop system is coerced into in between switches is that, for highly nonlinear uncertain systems, it may lead to finite escape time instability.

In a different vein, closed loop performance may be improved using the originally proposed cost-detectable cost function (without signal filtering), if other techniques are used to shape the response between the switches. In particular, neural networks have long been known as powerful tools for approximating unknown nonlinear functions. This fact extends their application to the control of unknown, nonlinear dynamic systems. In [20], Cao and Stefanovic proposed a combination of the neural networks and safe switching control, and showed that it led to improved performance. As an example, the candidate controller set is chosen to have a proportional integral derivative (PID) structure, where the parameters of the PID controllers are adaptively tuned via switching according to the hysteresis Algorithm 2.1 and a cost detectable performance index. This ensures that the closed loop system is stable, and that in addition, any destabilizing controller will not stay in the loop unnecessarily for long; as soon as the unstable modes get excited, the algorithm discards the controller and replaces it with an as yet unfalsified one (according to the feasibility assumption). However, there is still room for improvements in the dynamic performance in between any two switching time instants. It has been proposed in [54], that when a new PID controller is selected, the switching control algorithm resets the states of the integrator term, and the approximate differentiator term to prevent the discontinuity. To further improve the performance, the principle of radial basis function neural networks (RBFNN) can be used to update the parameters of the selected controller.

### 3.3.1 *Radial Basis Function Neural Networks in Safe Switching Control*

Consider the safe switching control algorithm with the PID candidate controller set. The conventional PID controllers with fixed parameters may deteriorate the control performance in accordance with the complexity of the plant. Hence, once the best optimal controller is selected using Algorithm 2.1 and the cost function (3.3), a three-layer RBFNN (Figure 3.15) is activated, to adaptively update the controller parameters to achieve better performance.

In the RBFNN, $X = [x_1, x_2, ...x_n]^T$ denotes the input vector. Assume that the radial basis vector is $H = [h_1, h_2, ...h_j, ...h_m]^T$, where $h_j$ is the Gaussian function:

$$h_j = \exp\left(-\frac{\|X - C_j\|^2}{2b_j^2}\right) \quad (j = 1, 2, ..., m) \tag{3.12}$$

and $C_j = \left[c_{j1}, c_{j2}, ...c_{ji}, ..., c_{jn}\right]^T$ is the center vector of the $j^{th}$ node.

Let $B = [b_1, b_2, ..., b_m]^T$, $b_j > 0$ be the basis width of the $j^{th}$ node; let $W = [w_1, w_2, ..., w_j, ...w_m]^T$ be the weight vector. As shown in Figure 3.15, the neural network output is

$$y_m = \sum_{j=1}^{m} w_j h_j. \tag{3.13}$$

The RBFNN identification performance index is chosen as

$$E = \frac{1}{2}\left(y(k) - y_m(k)\right)^2. \tag{3.14}$$

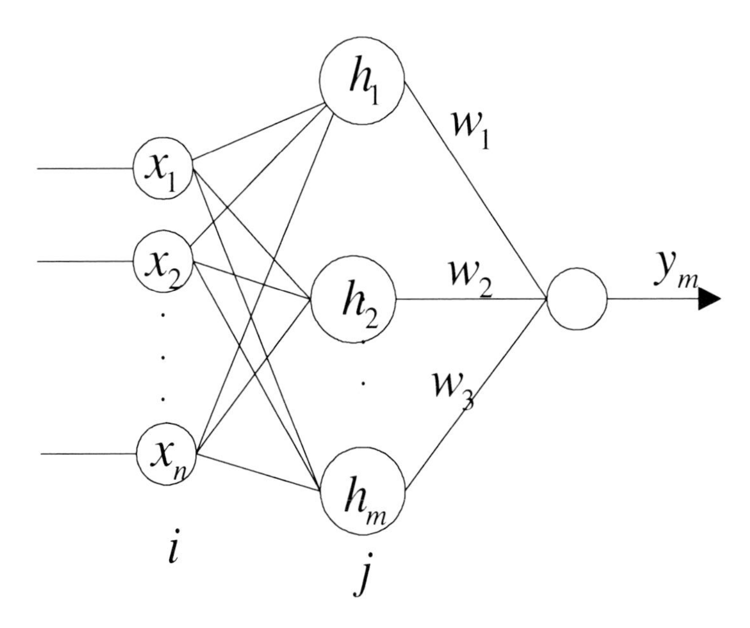


**Fig. 3.15** RBF neural network Structure

Then, according to the gradient descent algorithm, the output weights, node centers, and basis width parameters are calculated as follows:

$$\begin{aligned}
\Delta w_j &= \frac{\partial E}{\partial w_j} = (y(k) - y_m(k))h_j, \\
w_j(k) &= w_j(k-1) + \eta \Delta w_j + \alpha(w_j(k-1) - w_j(k-2)), \\
\Delta b_j &= \frac{\partial E}{\partial b_j} = (y(k) - y_m(k))w_j h_j \frac{\|X - C_j\|^2}{b_j^3}, \\
b_j(k) &= b_j(k-1) + \eta \Delta b_j + \alpha(b_j(k-1) - b_j(k-2)), \\
\Delta c_{ji} &= \frac{\partial E}{\partial c_j} = (y(k) - y_m(k))w_j h_j \frac{x - c_{ji}}{b_j^2}, \\
c_{ji}(k) &= c_{ji}(k-1) + \eta \Delta c_{ji} + \alpha(c_{ji}(k-1) - c_{ji}(k-2))
\end{aligned} \tag{3.15}$$

where $\eta$ is the learning rate and $\alpha$ is the momentum factor.

The inputs of the RBFNN identification are chosen as

$$\begin{aligned}
x_1 &= \Delta u(k) = u(k) - u(k-1), \\
x_2 &= y(k), \\
x_3 &= y(k-1)\,.
\end{aligned} \tag{3.16}$$

The Jacobian algorithm, used in the sequel, is given by:

$$\frac{\partial y(k)}{\partial \Delta u(k)} \approx \frac{\partial y_m(k)}{\partial \Delta u(k)} = \sum_{j=1}^{k} w_j h_j \frac{c_{ji} - \Delta u(k)}{b_j^2} . \tag{3.17}$$

Here, we use this algorithm to update the PID controller parameters, selected via switching using Algorithm 2.1. Define the remaining error of the unfalsified control as

$$error(k) = r - y(k) . \tag{3.18}$$

Hence, the inputs of the RBFNN based PID controller are

$$\begin{aligned} error\_p &= error(k) - error(k-1), \\ error\_i &= error(k), \\ error\_d &= error(k) - 2error(k-1) + error(k-2) . \end{aligned} \tag{3.19}$$

The control algorithm is given as:

$$\begin{aligned} u(k) &= u(k-1) + \Delta u(k), \\ \Delta u(k) &= k_p(error(k) - error(k-1)) + k_i(error(k)) + \\ &\quad + k_d(error(k) - 2error(k-1) + error(k-2)) . \end{aligned} \tag{3.20}$$

Here, the neural network approximation index is

$$E(k) = \frac{1}{2} error(k)^2 \tag{3.21}$$

where $k_p, k_i$, and $k_d$ are adjusted by the gradient descent algorithm:

$$\begin{aligned} \Delta k_p &= -\eta \frac{\partial E}{\partial k_p} = -\eta \frac{\partial E}{\partial y} \frac{\partial y}{\partial \Delta u} \frac{\partial \Delta u}{\partial k_p} = \eta error(k) \frac{\partial y}{\partial \Delta u} error\_p, \\ \Delta k_i &= -\eta \frac{\partial E}{\partial k_i} = -\eta \frac{\partial E}{\partial y} \frac{\partial y}{\partial \Delta u} \frac{\partial \Delta u}{\partial k_i} = \eta error(k) \frac{\partial y}{\partial \Delta u} error\_i, \\ \Delta k_d &= -\eta \frac{\partial E}{\partial k_d} = -\eta \frac{\partial E}{\partial y} \frac{\partial y}{\partial \Delta u} \frac{\partial \Delta u}{\partial k_d} = \eta error(k) \frac{\partial y}{\partial \Delta u} error\_d \end{aligned} \tag{3.22}$$

where $\frac{\partial y}{\partial \Delta u}$ is the plant Jacobian, which is calculated by the RBFNN.

The control structure is shown in Figure 3.16. According to [54], when a new controller is selected, the control algorithm resets the states of the integrator term, and the approximate differentiator term (to prevent the discontinuity). Thus, the entire algorithm can be presented as follows. The data $u$ and $y$ are measured; $\tilde{r}_i$ and $\tilde{e}_i$ are calculated for each candidate controller and the measured plant data; $J_i$ is calculated, and the controller $\arg \min_{1 \le i \le N} J_i(t)$ is switched into the loop. At the switching times, the controller states are reset, and the RBFNN is combined with the selected

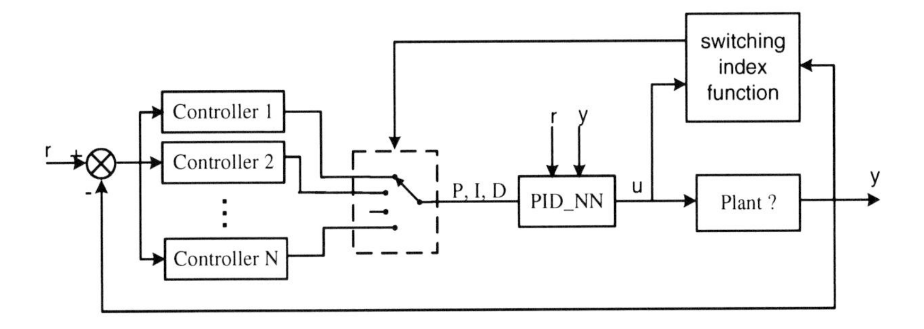


**Fig. 3.16** Switching Control Using NN

controller to update the PID parameters. The combination of neural networks with the switching adaptive control does not alter the stability of the switched system. Once the RBFNN is combined with the selected controller, the switched controller remains in the candidate controller set, in which all controllers are supervised by the switching law. When the measured data start revealing instability, the currently active controller is quickly switched out of the loop and replaced by another, as yet unfalsified one. Hence, stability of the overall switched unfalsified control system combined with the neural networks is preserved (under the feasibility assumption).

In the following, MATLAB® simulation results of [20] are presented, with zero disturbance and noise, and zero initial conditions, though this can be relaxed.

To illustrate the algorithm described above and compare with the simulation results of multiple model adaptive control in [10], the same simulation setting as in [10] is reproduced. The transfer function of the actual plant (unknown from the control perspective) is assumed to be:

$$G_p(s) = \frac{a}{s^2 + bs + c} \,. \tag{3.23}$$

The parameters $a, b$, and $c$ are assumed to lie in the compact set given as $S = \{0.5 \leq a \leq 2, -0.6 \leq b \leq 3.4, -2 \leq c \leq 2\}$. The reference input is a square wave signal with a unit amplitude and a period of 10 seconds, and the reference model, whose output is to be tracked, is taken as $W_m(s) = \frac{1}{s^2+1.4s+1}$. Simulations were conducted on the three plants suggested in [10], all unstable oscillatory plants: $\frac{0.5}{s^2-0.35s+2}$, $\frac{0.5}{s^2-0.5s+2}$, and $\frac{0.5}{s^2+0.5s-2}$.

*Simulation* 1 : Here it is demonstrated that the cost detectable cost function is "safer" than the non cost detectable one. In this case, the controller parameter set is taken as $K_P = \{1,5,20,50,100\}$, $K_I = \{1,5,20,30,50,100\}$, and $K_D = \{0.2,0.5,1,5,15\}$. The simulation results are shown in Figures 3.17, 3.19 and 3.20, where column (a) represents the reference input $r(t)$, the reference model output $y_m(t)$, the plant

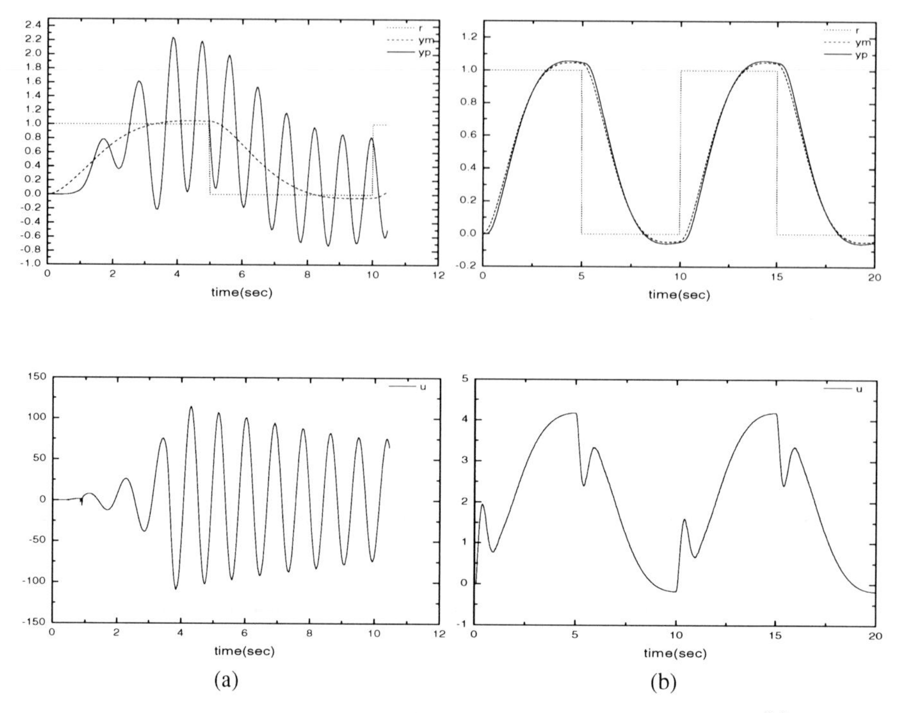


**Fig. 3.17** Input/output (I/O) data for the unstable oscillatory plant $G_p(s) = \frac{0.5}{s^2-0.35s+2}$ with two different cost functions

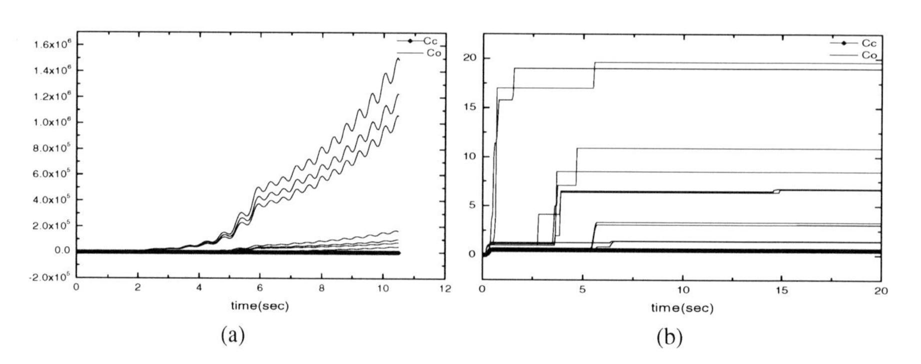


**Fig. 3.18** Controller cost with two different cost functions, with the plant $G_p(s) = \frac{0.5}{s^2-0.35s+2}$

response $y_p(t)$ and plant input $u(t)$ with the non cost detectable cost function ($J_i(t) = -\rho + \int_0^t \Gamma_{spec}(\widetilde{r_i}(t), y(t), u(t))dt$), where $\rho > 0$ is used to judge whether a certain controller is falsified or not, and $\Gamma_{spec}$ is chosen as

$$\Gamma_{spec}(\widetilde{r_i}(t), y(t), u(t)) = (w_1 * (\widetilde{r_i}(t) - y(t))^2 + (w_2 * u(t))^2 - \sigma^2 - \widetilde{r_i}(t)^2 \quad (3.24)$$

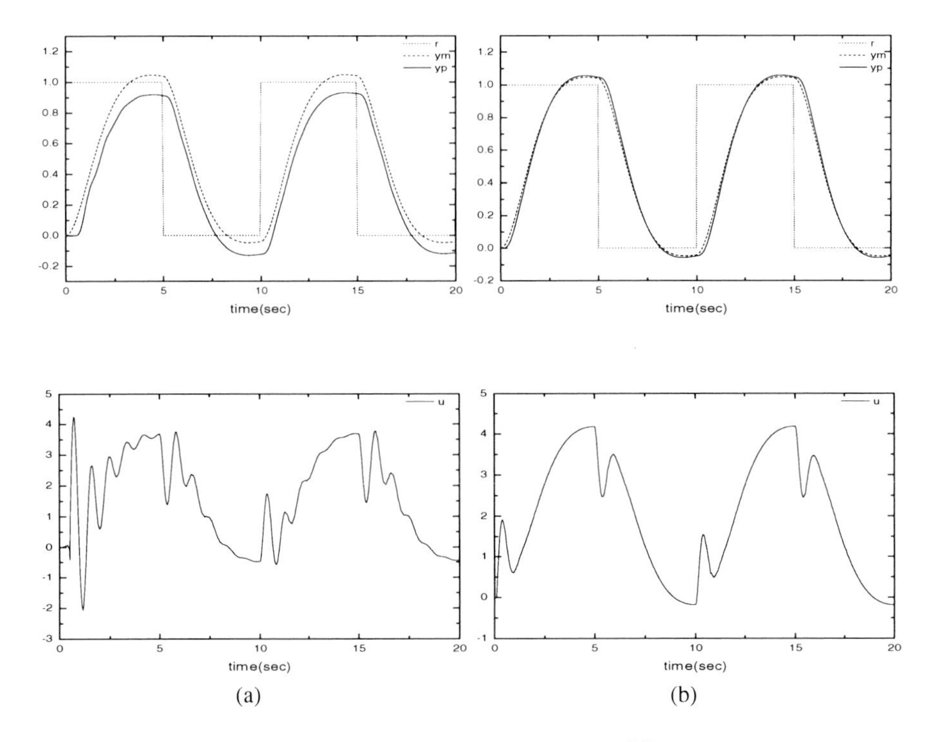


**Fig. 3.19** I/O data for the unstable oscillatory plant $G_p(s) = \frac{0.5}{s^2-0.5s+2}$ with two different cost functions

where $w_1$ and $w_2$ are the weighting filters chosen by the designer, and $\sigma$ is a constant representing the root-mean-square (r.m.s.) effects of noise on the cost. The column (b) shows the corresponding results with the cost detectable cost function( $J_i(t) = -\rho + \max_{\tau\in[0,t]} \frac{\|u\|_\tau^2 + \left\|\tilde{e}_i\right\|_\tau^2}{\left\|\tilde{r}_i\right\|_\tau^2}$). In Figures 3.17 (a), 3.20 (a), the outputs $y_p(t)$ are both oscillatory divergent, and all the candidate controllers are falsified in approximately 11 seconds. In Figure 3.19 (a), $y_p(t)$ is bounded, but the tracking error is rather large. In Figures 3.17, 3.19 and 3.20 (b), the tracking error is considerably smaller. To clarify the difference between the above two cost functions, we decreased the dimension of controller set to $K_P = \{1, 50, 100\}$, $K_I = \{1, 50\}$, and $K_D = \{0.2, 5, 15\}$. Figure 3.18 (a) shows the cost with the non cost detectable cost function with the plant $G_p(s) = \frac{0.5}{s^2-0.35s+2}$, while Figure 3.18 (b) represents the corresponding cost with the cost detectable cost function. In Figure 3.18, lines $Cc$ and $Co$ respectively stand for the cost of the selected (active) controller and the costs of the non-selected ones. It is seen that the former cost function lacks the capability of selecting the "best" controller, or they even discard the stabilizing controller. As expected, the stability and performance with the latter cost function are preserved. Moreover, compared with the results in [10], we can conclude that the safe switching control, under the

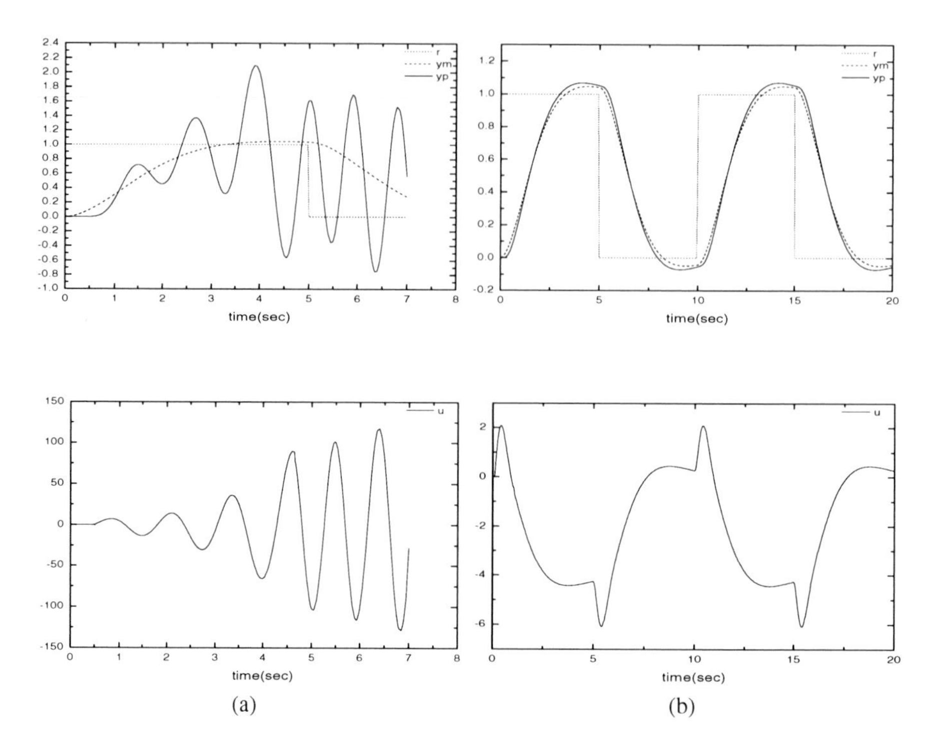


**Fig. 3.20** I/O data for the unstable nonoscillatory plant $G_p(s) = \frac{0.5}{s^2+0.5s-2}$ with two different cost functions

feasibility assumption and with a properly designed cost function, can achieve similar, or even better response results.

*Simulation* 2 : The effectiveness of neural networks combined with the unfalsified control is demonstrated here. In this case, the controller set is set to be smaller to make results more obvious, e.g. $K_P = \{5, 50, 100\}$, $K_I = \{20, 30\}$, and $K_D = \{0.2, 0.5, 1\}$. The cost function is the chosen to be the cost detectable one. Other simulation parameters are the same as in the previous setting. Figures 3.21, 3.23, and 3.24 (b) show the simulation results with neural networks engaged to update the PID parameters of the currently active controller. Figures 3.21, 3.23, and 3.24 (a) show the simulations without neural networks. By comparing Figures 3.21, 3.23, and 3.24 (b) with Figures 3.21, 3.23, and 3.24 (a), we can find that the combination of neural networks with the safe switching control reduces the tracking error, and even stabilizes a divergent system, in the scenarios when the candidate controller set is small or does not match well with the actual plant, and all controllers in the controller set might be falsified. Figure 3.22 shows the PID gains, which demonstrate the effect of neural networks.

Although the tracking error is increasing in the scenarios without neural networks (Figures 3.21, 3.23, and 3.24 (a)), as soon as stability is falsified by the switching

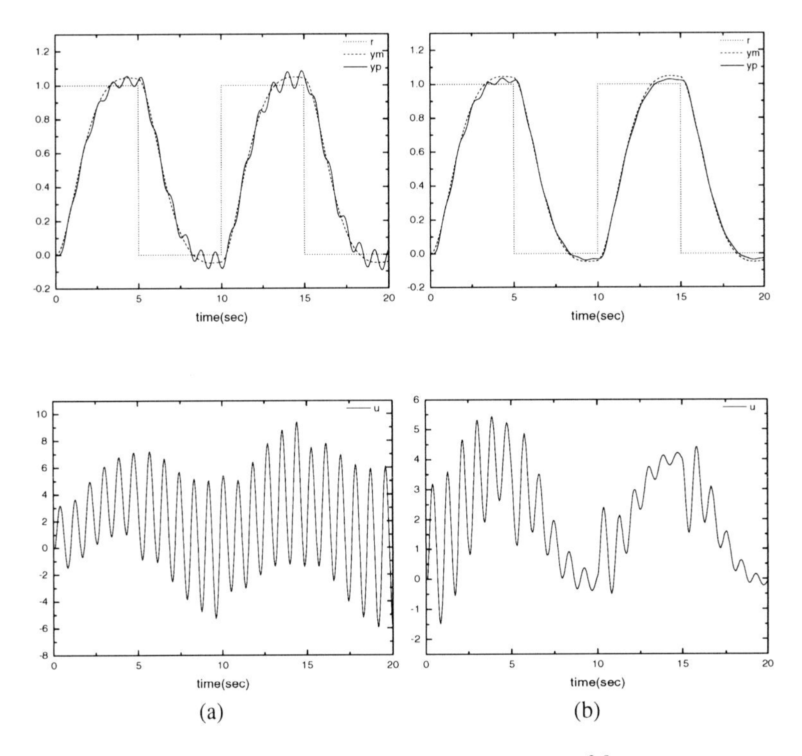


**Fig. 3.21** I/O data for the unstable oscillatory plant $G_p(s) = \frac{0.5}{s^2 - 0.35s + 2}$ $(a)$ without neural networks, $(b)$ with neural networks

algorithm (guaranteed by the cost detectable cost function), the currently active destabilizing controller will be switched out and replaced with an as yet unfalsified one. However, every new switching would require controller state resetting to prevent adverse transients in the output, which is circumvented by the proper use of neural networks, as demonstrated.

### 3.3.2 Bumpless Transfer with Slow-Fast Controller Decomposition

An alternative way of reducing the discontinuities and abrupt transients during the switching times has been considered in [23]. It is an extension of the bumpless transfer methods that have been investigated since the 1980s (see, for example, [39], [102], [116]). In adaptive control however, the plant is not precisely known at the outset, and the goal of adaptive control is to change the controllers to improve performance as the plant data begin to reveal some information about the plant. Thus, in adaptive switching control an exact plant is generally unavailable at the time of switching. This implies that bumpless transfer methods that may be suitable for non-adaptive applications such like anti-windup or transfer from manual to automatic

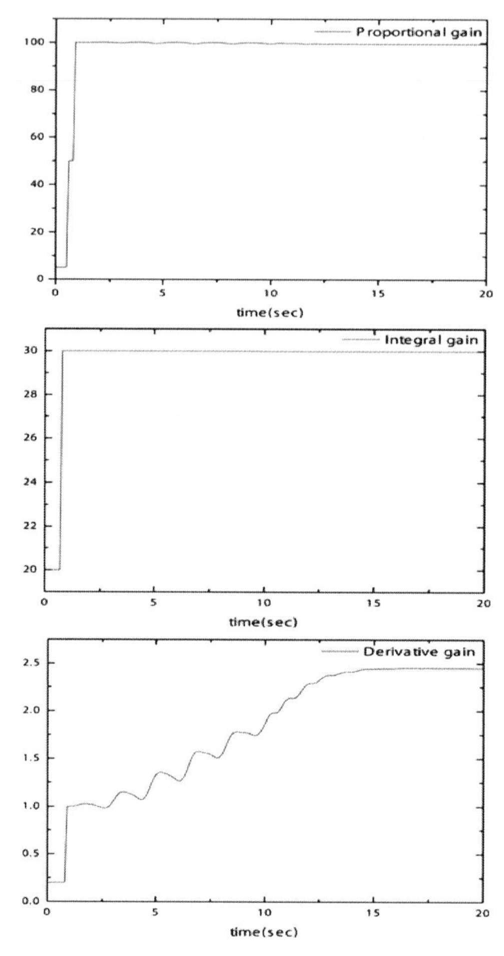


**Fig. 3.22** PID gains for the neural network augmented system for the unstable oscillatory plant $G_p(s) = \frac{0.5}{s^2-0.35s+2}$

control where the true plant is well-known, may not be ideal for adaptive switching control applications. In particular, in adaptive switching applications where the true plant model may only be poorly known at the controller switching times, it may be preferable to employ a bumpless transfer technique for adaptive control that does not depend on precise knowledge of the true plant model. In contrast to the model-based bumpless transfer methods [39], [116], the continuous switching methods of [5], conditioning methods of [40] and linear quadratic optimal bumpless transfer of [102] do not require a plant model.

The exposition in [23] is based on a slow-fast decomposition of the controller. Similarly to [5], it does not require precise knowledge of the plant at switching times. This kind of decomposition approach is inspired by the adaptive PID

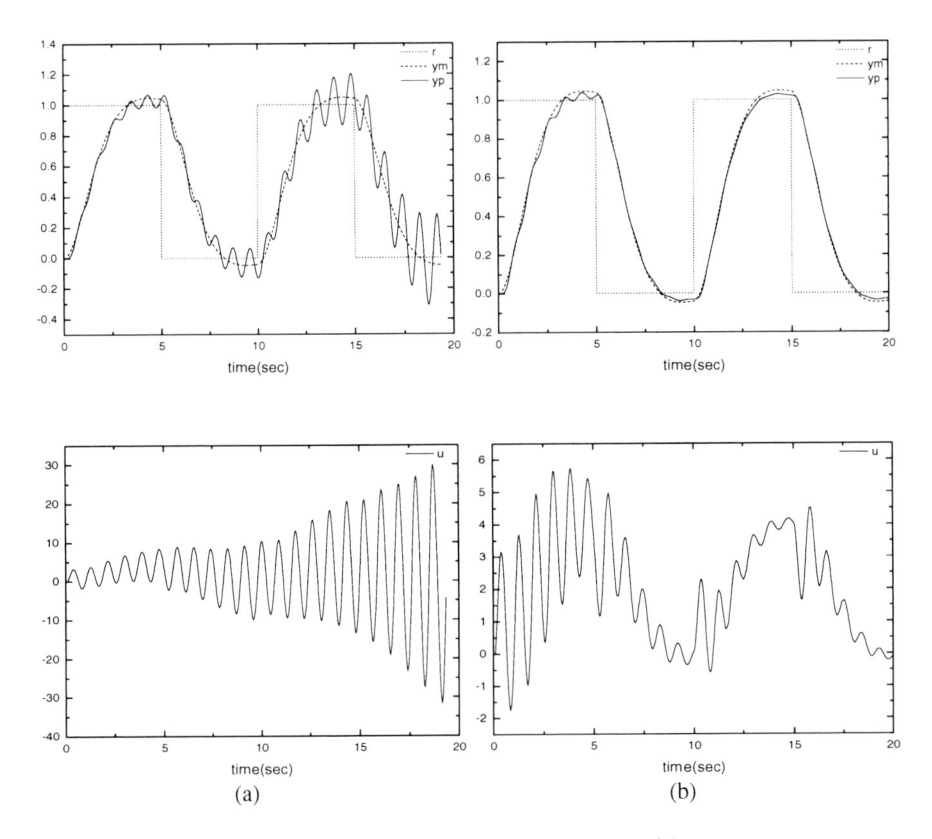


**Fig. 3.23** I/O data for the unstable oscillatory plant $G_p(s) = \frac{0.5}{s^2-0.5s+2}$ $(a)$ without neural networks, $(b)$ with neural networks

controller proposed in [54]. PID controller is a special case of a controller with fast modes (the differentiator) and slow modes (the integrator). Generalizing the PID controller case, the bumpless transfer suggested in this chapter decomposes the original controllers into the fast mode controllers and the slow mode controllers. By appropriately re-initializing the states of the slow and fast modes at switching times, the method in [23] can ensure not just the controller output continuity, but also avoidance of the fast transient bumps after switching. We present these results in the following.

Consider a switching control system shown in Figure 3.25. The system includes a plant and a set of 2-DOF controllers

$$\mathbf{K} = \{K_1, ..., K_i, ..., K_n\} \quad (i = 1, 2, ..., n)\,. \tag{3.25}$$

Assume that the plant output is continuous when input is continuous; a linear time invariant plant with a proper transfer function is a good example. The input of the

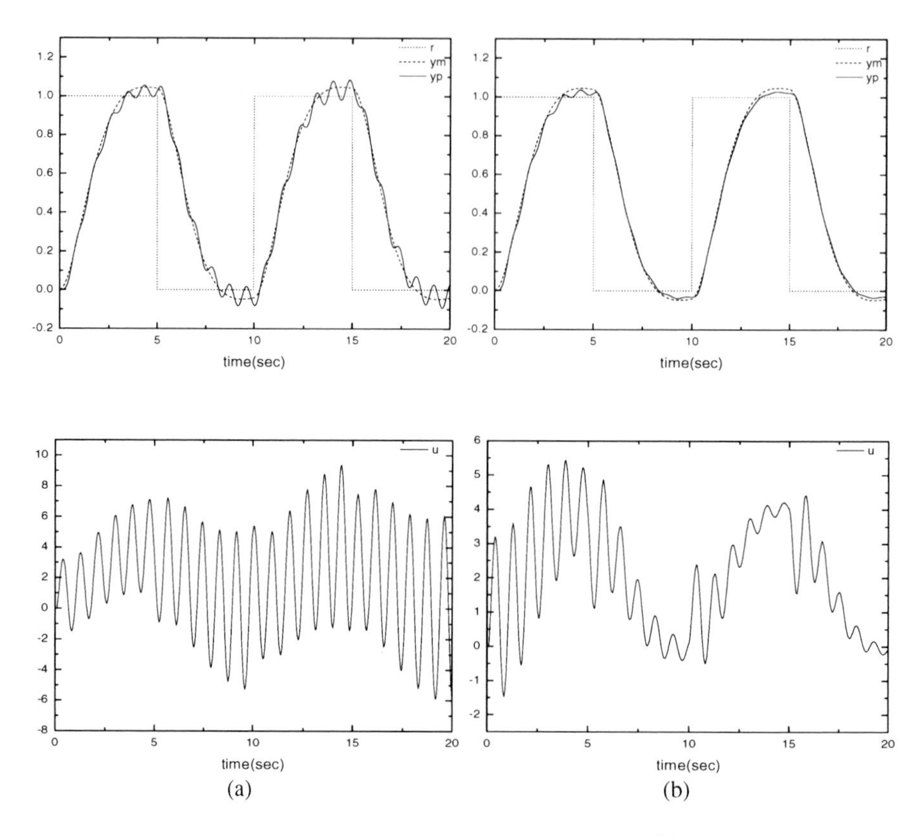


**Fig. 3.24** I/O data for the unstable nonoscillatory plant $G_p(s) = \frac{0.5}{s^2+0.5s-2}$ $(a)$ without neural networks, $(b)$ with neural networks

plant is $u(t)$ and the output is $y(t)$. The plant input is directly connected to the controller output. Controller inputs are $r(t)$ and $y(t)$ where $r(t)$ is a reference signal.

When a controller $K_i$ is in the feedback loop, the controller is said to be on-line, and the other controllers are said to be off-line. The $i^{th}$ controller $K_i$ is supposed to have state-space realization

$$\begin{aligned}\dot{x}_i &= A_i x_i + B_i z,\\ y_{K_i} &= C_i x_i + D_i z,\end{aligned}$$

where $z = \left[r^T\ y^T\right]^T$ is the input and $y_{K_i}$ is the output of $K_i$. Equivalently, one can write

$$K_i(s) \triangleq \left[\begin{array}{c|c} A_i & B_i \\ \hline C_i & D_i \end{array}\right].$$

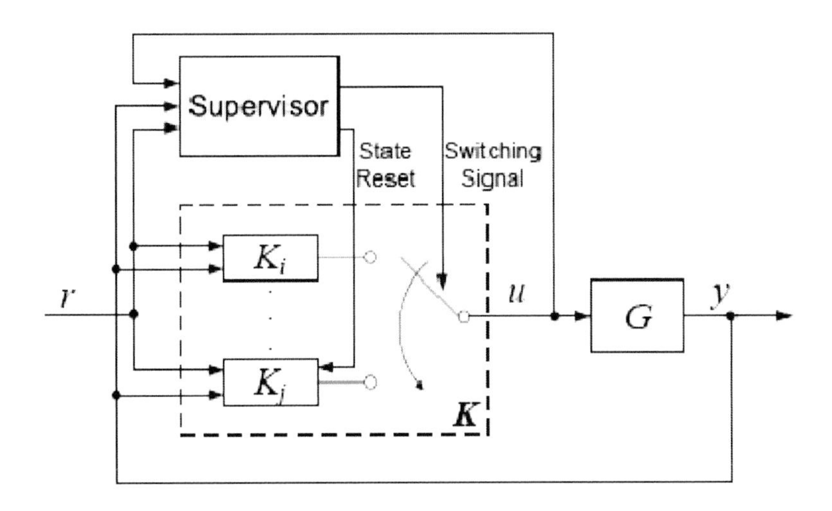


**Fig. 3.25** Switching control system with two-degree-of-freedom controllers

When the on-line controller is switched from $K_i$ to $K_j$ at time $t_s$, the control input can be written as

$$u = \left\{ \begin{array}{ll} y_{K_i}, & for\ t < t_s \\ y_{K_j}, & for\ t \geq t_s \end{array} \right\} \tag{3.26}$$

where the switching time (or switching instant) is denoted by $t_s$.

Since the controller output $y_{K_i}$ is replaced by $y_{K_j}$ at the switching instant $t_s$, the control signal can have bumps in the neighborhood of $t = t_s$ if $y_{K_i}$ and $y_{K_j}$ have different values. In the following, the times immediately before and after $t_s$ are denoted $t_s^-$ and $t_s^+$, respectively. The objective of the bumpless transfer is to ensure continuity at the switching instant, and immediately following it.

**Slow-fast decomposition**

Consider the following decomposition of the controllers into slow and fast parts:

$$K(s) = K_{slow}(s) + K_{fast}(s) \tag{3.27}$$

with respective minimal realizations

$$K_{slow}(s) \triangleq \left[ \begin{array}{c|c} A_s & B_s \\ \hline C_s & D_s \end{array} \right]$$

and

$$K_{fast}(s) \triangleq \left[ \begin{array}{c|c} A_f & B_f \\ \hline C_f & D_f \end{array} \right] .$$

The poles of the slow portion $K_{slow}(s)$ are of smaller magnitude than the poles of $K_{fast}(s)$, *i.e.*,

$$|\lambda_i(A_s)| \leq |\lambda_j(A_f)| \; for\ all\ i,j$$

where $\lambda(\cdot)$ denotes the $i^{th}$ eigenvalue.

Note that the $K_{slow}(s)$ and $K_{fast}(s)$ portions of the slow-fast decomposition may be computed by various means, *e.g.*, the MATLAB® `slowfast` algorithm, which is based on the stable-antistable decomposition algorithm described in [89].

The slow-fast decomposition of the $i^{th}$ controller $K_i$ in the set $K$ is denoted with the subscript $i$:

$$K_{islow}(s) \triangleq \left[\begin{array}{c|c} A_{is} & B_{is} \\ \hline C_{is} & D_{is} \end{array}\right] \tag{3.28}$$

and

$$K_{ifast}(s) \triangleq \left[\begin{array}{c|c} A_{if} & B_{if} \\ \hline C_{if} & D_{if} \end{array}\right] . \tag{3.29}$$

Now, the bumpless transfer problem is defined in [23] as follows:

**Definition 3.1.** A switching controller with slow-fast decomposition (3.27) is said to perform a *bumpless transfer* if, whenever a controller is switched, the new controller state is reset so as to satisfy bith of the following two conditions:

1. The control input signal $u(t)$ is continuous at $t_s$ whenever $r_t \in C^0$, and
2. The state of the fast part of the controller ($K_{fast}(s)$ is reset to zero at $t_s$.

The condition (1) in Definition 3.1 can be found in other bumpless transfer literature, as well. The condition 2 is concerned with the control signal after switching. This additional requirement for the bumpless transfer is needed, as argued in [23], to insure that there are no rapid transients immediately following controller switching.

The proposed bumpless transfer method with slow-fast decomposition is based on the following assumption for each candidate controller:

**Assumption 3.1.** For each candidate controller $K_i$, the slow part $K_{islow}$ in (3.28) has at least $m = dim(u)$ states.

The idea can be explained as follows. The Assumption 3.1 is sufficient to allow the state of the slow controller $K_{islow}$ to be reset at switching times to ensure both continuity and smoothness of the control signal $u(t)$. In general, even if all the controllers have the same order and share a common state vector, when the controller switching occurs, any or all of the slow and fast controller state-space matrices will be switched, which can lead to bumpy transients or discontinuity in the control signal $u(t)$ at switching times. However, if only $A_{is}$ or $B_{is}$ are switched and there is a common state vector before and after the switch, then the control signal will be continuous and no bumpy fast modes of the controller will be excited. Fast transient bumps or discontinuities, when they occur, may arise from switching the $D_{is}$ matrix of the slow controller or from switching any of the state-space matrices $(A_{if}, B_{if}, C_{if}, D_{if})$ of the fast controller. In the case of switching the matrices $A_{if}$ or $B_{if}$, switches do not actually result in discontinuous jumps in $u(t)$, but nevertheless can result in the bumpy fast transients in the control signal which, if very fast, may appear to be nearly discontinuous.

The goal in bumpless transfer is to avoid both discontinuity and fast transients induced by changing fast modes. The method should work even when the order of the controller changes at the switching times, and to allow for the possibility that the true plant may be imprecisely known, it is preferred that the switching algorithm does not depend on precise knowledge of the true plant. In this method, it can be done by initializing the state of the slow part of the new controller $K_{jslow}(s)$ after each switch to a value computed to ensure continuity, and setting the state of the fast part $K_{jfast}(s)$ to zero.

**Theorem 3.1.** *Suppose that each of the candidate controllers have slow-fast decomposition (3.28) and (3.29) satisfying Assumption 3.1, and suppose that at time $t_s$ the online controller is switched from controller $K_i$ to controller $K_j$. At $t_s$, let the states of the slow and fast controllers be reset according to the following algorithm:*

**Algorithm 3.1. Slow-fast bumpless transfer algorithm**

$$x_{fast}(t_s^+) = 0, \tag{3.30}$$

$$x_{slow}(t_s^+) = C_{js}^{\dagger}\{u(t_s^-)(D_{js} + D_{jf})z(t_s^-)\} + \xi, \tag{3.31}$$

*where* $z = \left[r^T, y^T\right]^T$, $C_{js}^{\dagger}$ *is the pseudoinverse matrix of* $C_{js}$, *and* $\xi$ *is any element of the null space of* $C_{js}$*:*

$$C_{js}\xi = 0\,. \tag{3.32}$$

*Then, bumpless transfer is achieved at the switching time $t_s$.*

*Proof.* The control signal immediately after switching (time $t_s^+$) can be written, based on state space representation model (3.28) and (3.29) of the new controller $K_j(s)$, as

$$u(t_s^+) = C_{js}x_{slow}(t_s^+) + C_{jf}x_{fast}(t_s^+) + (D_{js} + D_{jf})z(t_s^+)\,. \tag{3.33}$$

The equations 3.30 and 3.31 imply:

$$u(t_s^+) = C_{js}\left[C_{js}^{\dagger}\{u(t_s^-) - (D_{js} + D_{jf})z(t_s^-)\} + \xi\right] + (D_{js} + D_{jf})z(t_s^+)\} \tag{3.34}$$

By Assumption 3.1, $C_{js}C_{js}^{\dagger} = I_{m\times m}$ where $m$ is larger than or equal to the number of states of $K_j$. This results in:

$$u(t_s^+) = u(t_s^-) - (D_{js} + D_{jf})z(t_s^-) + (D_{js} + D_{jf})z(t_s^+)\,. \tag{3.35}$$

Since

$$z(t_s^-) = \left[r^T(t_s^-)\, y^T(t_s^-)\right]^T = \left[r^T(t_s^+)\, y^T(t_s^+)\right]^T = z(t_s^+) \tag{3.36}$$

we have

$$u(t_s^-) = u(t_s^+). \tag{3.37}$$

The result follows immediately from the bumpless transfer definition.

Note that since $C_{js}$ is a full rank matrix which consists of $m$ linearly independent vectors, $C_{js}C_{js}^T$ is invertible and the following holds:

$$C_{js}^\dagger = C_{js}^T(C_{js}C_{js}^T)^{-1} \,.$$

The slow-mode controllers that satisfy Theorem 3.1 are constructed as follows. Using observable canonical form of $K_{slow}$ is a good choice since $C_{js}$ in (3.33) is of the form

$$C_{js} = [I\ 0\ \ldots\ 0] \tag{3.38}$$

in which case $C_{js}^\dagger = C_{js}^T$.

Since (3.38) holds for all $j$, one uses transpose matrices rather than pseudoinverse matrices of $C_{js}, \forall j$. This reduces the complexity of the state reset procedure.

Note that the bumpless transfer in [5] is a special case of the bumpless transfer method in [23]. The method in [5] requires that all controllers have the same number of states and all controllers have a state-space realization sharing a common $C$-matrix and $D$-matrix. The slowfast method in [23] does not impose these controller restrictions, and can be used whenever the minimal realization of the slow part $K_{islow}$ has order at least equal to the $dim(u)$. However, if the controllers have a slow part only (*i.e.*, $K_{ifast} = 0$, $\forall i$) and each controller has $C$-matrices of the form of (3.38) and a common $D$-matrix (*e.g.* $D_{is} = 0$, $\forall i$), then it is expected that both methods will have the same result. The advantages of the slow-fast bumpless transfer method of [23] arise when the controllers have both slow and fast modes, in which case the latter method is able to exploit the additional flexibility for state re-initialization provided by the additional fast modes to eliminate the bumpy abrupt transients that might otherwise result.

**Controllers with single type of modes**

The cases when some candidate controllers have slow modes only or fast modes only can be addressed as special cases of the slow-fast decomposition bumpless transfer. If candidate controllers have slow modes only, then using the observable canonical form and applying (3.31) in Theorem 3.1 solves the problem. On the other hand, when the candidate controllers have fast modes only $K = K_{fast}$, it is necessary to modify the controllers appropriately to apply Theorem 3.1 because they do not contain any slow parts to be re-initialized. One possible solution is augmenting the controller states with uncontrollable slow modes that were not originally contained in the controller. Then, by slow-fast decomposition, an additive slow mode controller $K_{slow}$ is included in the controller so that $\tilde{K} = K_{slow} + K_{fast}$ where

$$K_{slow}(s) \triangleq \left[\begin{array}{c|c} A_s & 0 \\ \hline C_s & 0 \end{array}\right]$$

and $A_s \in R^{m \times n}$ has only slow modes. Note that the $K_{slow}$ has zero matrices for its $B_s$ and $D_s$, so its output is determined solely by the initial state. The matrix $C_s$ is written in the form of (3.38) for simplicity. Now, (3.30) and (3.31) in Theorem 3.1 can be applied, similarly as was applied when there was already a $K_{slow}$ part. Since $K(s) = \tilde{K}(s)$, the measurements for performance are not affected by adding the slow mode controller (3.3.2), except for transients after the switching times.

**Simulation results**

A reduction of abrupt transients using the above slow-fast bumpless transfer method is demonstrated through the following example in [23]. A comparison is shown below, among non-bumpless, continuity-assuring bumpless, and slow-fast decomposition bumpless transfers. By way of this example, the importance of the additional condition 2 in Definition 3.1 is shown, compared with a previously existing method in [5]. In the following realization, a PID controller has an infinitely slow pole and a very fast zero when $\varepsilon \neq 0$:

$$K(s) = K_{slow}(s) + K_{fast}(s) = K_P + K_I/s + \frac{K_D s}{\varepsilon s + 1}.$$

Since a fast zero in the differentiator part can make a large and fast transient even after the switch has occurred, considering only continuity of controller output as in [5] might not be sufficient to perform bumpless transfer. The method in [23] can, however, suppress adverse transients right after the switching, provided that fast and slow mode controllers are properly initialized. Consider the following plant:

$$G(s) = \frac{s^2 + s + 10}{s^3 + s^2 + 98s - 100}.$$

The simulation results [23] are shown next for the case of two controllers having the structure as in 3.26. The controller gains are chosen as follows:

$$Controller\ 1: K_{P1} = 80; K_{I1} = 50; K_{D1} = 0.5$$
$$Controller\ 2: K_{P2} = 5; K_{I2} = 2; K_{D2} = 1.25\ .$$

A small number $\varepsilon = 0.01$ and the reference input $r = 1$ are used in the simulation. The constant $\varepsilon$ prevents the differentiator from making an infinite peak when a discontinuity comes into the controller. A PID controller is naturally decomposed into a slow and a fast part. Since a proportional gain is a memoryless component, it can be added to either part. Controller input is $z = \left[r^T\ y^T\right]^T$ for 2-DOF controllers. Subsequently, the controllers were decomposed into

$$K_{slow}(s) \triangleq \left[\begin{array}{c|c} 0 & [K_I \; -K_I] \\ \hline 1 & [K_P \; -K_P] \end{array}\right] .$$

And, in the same way, $K_{fast}$ can be written as

$$K_{fast}(s) \triangleq \left[\begin{array}{c|c} -1/\varepsilon & [1/\varepsilon \; -1/\varepsilon] \\ \hline -K_D/\varepsilon & [K_D/\varepsilon \; -K_D/\varepsilon] \end{array}\right] .$$

Controller 1 and Controller 2 in this particular case are, respectively,

$$K_1(s) = K_{1slow} + K_{1fast} = 80 + \frac{50}{s} + \frac{0.5s}{0.01s+1},$$
$$K_2(s) = K_{2slow} + K_{2fast} = 5 + \frac{2}{s} + \frac{1.25s}{0.01s+1} .$$

Controller $K_1(s)$ was designed to stabilize the plant, while $K_2(s)$ cannot stabilize it. In this experiment, $K_2(s)$ is the initial active controller. Thus, the plant was not stabilized at the early stage. After 2 seconds, the on-line controller was switched to $K_1(s)$.

The bumpless transfer method in [5] does not include any initializing or state reset procedure at the switching instants. Instead, it works allowing only those controllers for which there exist state-space realizations that share common $C$ and $D$ matrices; *i.e.*,

$$C_i = C_j \triangleq C \; and \; D_i = D_j \triangleq D \forall i \neq j .$$

Note that this is not possible in general, unless all the controllers have the same order and same $D$-matrices. Even though the slow-fast method of [23] does not have this restriction, the following simulation example includes this requirement to be able to directly compare the two different methods under the same conditions. In total, three simulation experiments were performed. First, switching without any bumpless transfer method was performed. The second simulation used the method of [5] and the third one used the slow-fast method of [23] based on Theorem 3.1. The upper part of Figure 3.27 shows the controller output. The solid line [23] of output $u(t)$ shows a smooth transient around the switching instant, while the dashed line [5] shows a fast transient after switching. The dotted line indicates the switching transient without bumpless transfer, which has extremely high peak value generated by the derivative controller. If $\varepsilon \to 0$, the peak value goes to infinity. Figure 3.28 shows $u(t)$ with the time axis magnified near the switching time. While the output without bumpless transfer has a discontinuity at the switching time $t = 2$ $s$, the outputs with the bumpless transfer (dashed and solid lines) show continuous transients. Note that the dashed line [5] satisfies Condition 1 in Definition 3.1, which coincides with the definition of bumpless transfer used in [5]. However, comparing with the solid line, the dashed line exhibits a fast bumpy transient after 2 seconds. It is excited by changing $K_{fast}$, which is a clearly different result from [23]. The resulting plant output $y(t)$ shown in the lower part of Figure 3.27 likewise exhibits an abrupt transient with the method of [5]. Both the control signal and the plant output are significantly smoother with the slow-fast bumpless transfer method of [23].

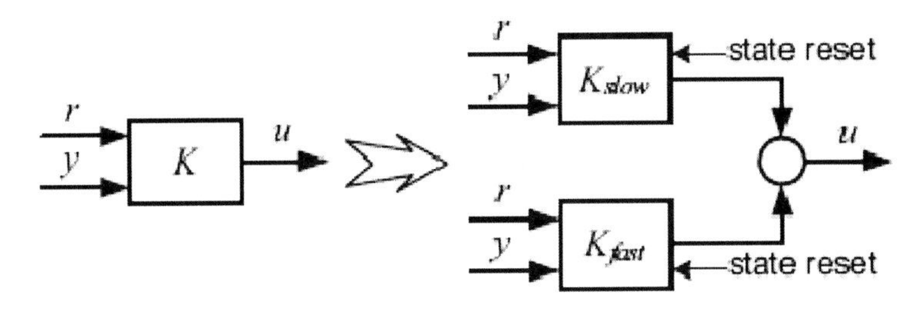


**Fig. 3.26** Slow-fast controller decomposition

## *3.3.3 Performance Improvement Using Bandpass Filters*

Some limitations of the traditional cost functions were pointed out in [26]: the standard $\mathfrak{L}_{2e}$-gain related cost function may allow a destabilizing controller to remain in the loop for an unnecessarily long time, during which the closed loop performance may deteriorate. To deal with this situation, an improvement by way of filtering the signals in the cost function with a series of bandpass filters is proposed in [22]. The performance improvement, inspired by Saeki *et al.* [87], is demonstrated by a faster convergence to a final controller.

In [26], Dehghani *et al.* discovered that an unfalsified controller designed using a traditional $\mathfrak{L}_{2e}$-gain related cost function could, for a low or zero dwell time, result in a destabilizing controller being placed into the loop for a long time relative to the longest time constant of the system. This produced adverse transients until another controller was settled on. The authors of [26] suggested that this indicates a limitation of the unfalsified adaptive control framework. Instead of focusing on the introduction of a dwell time favored by [26], the results of [22] showed that the original $\mathfrak{L}_{2e}$-gain related cost function may be slow to detect large and fast transients in the observed data in certain cases, and thus can take a long time to respond to evidence of unstable behavior in the data.

In [22], the following filtered cost function $V$ is proposed, where $F_j$ denotes a filter and $\mathbf{F}$ the set of available filters:

$$\begin{aligned} W(C_i,(u,u),\tau) &= \max_j \frac{||F_j u||_\tau^2 + ||F_j(\tilde{r}_i - y)||_\tau^2}{||F_j\tilde{r}_i||_\tau^2 + \alpha}, \\ V(C_i,(u,u),\tau) &= \max_{\tau\in[0,t]} W(C_i,(u,u),\tau)\,. \end{aligned} \tag{3.39}$$

Each filter's pass band is designed to cover a distinct frequency range with no frequency overlaps among the filters. In addition, the sum of all filters should cover the entire frequency range $\omega \in [0,\infty)$, and the sum of the squares of the filters should

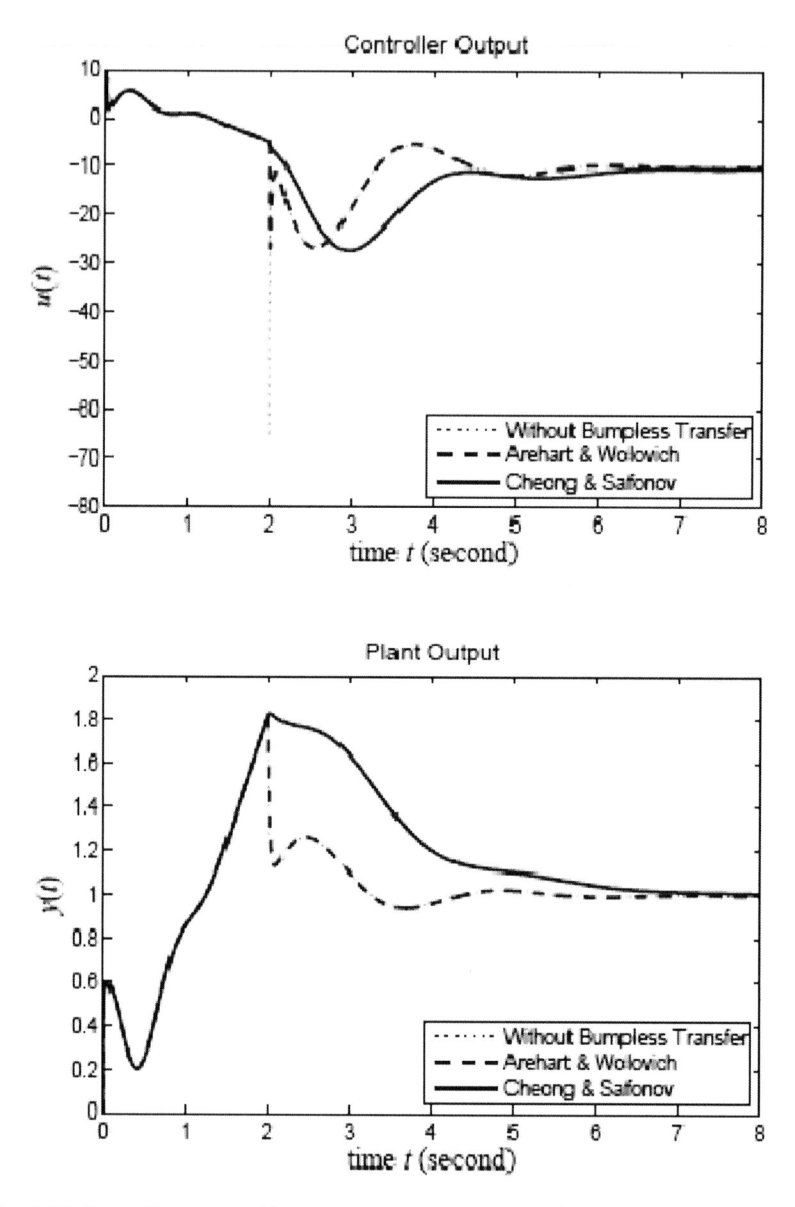


**Fig. 3.27** Controller output $u(t)$ (upper figure); plant output $y(t)$ (lower figure). Controller is switched at $t = 2$ $s$.

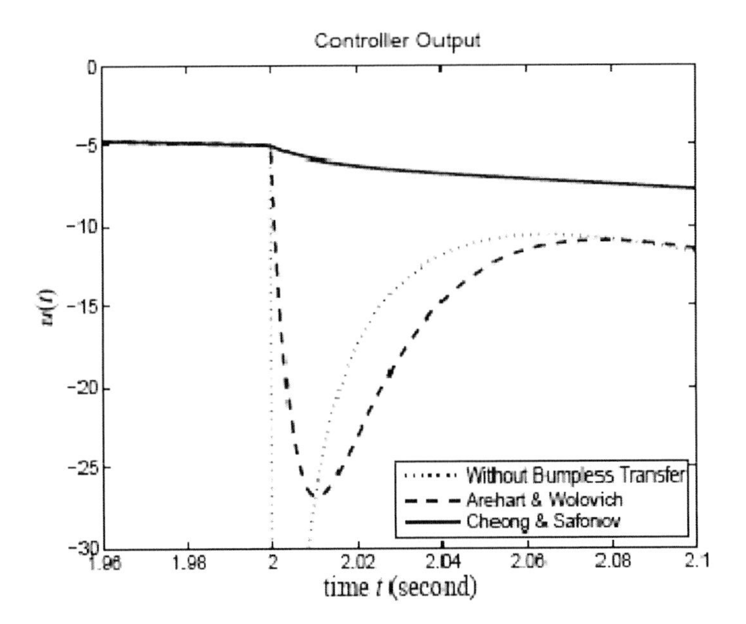


**Fig. 3.28** Magnified $u(t)$ around the switching instant ($t = 2\ s$)

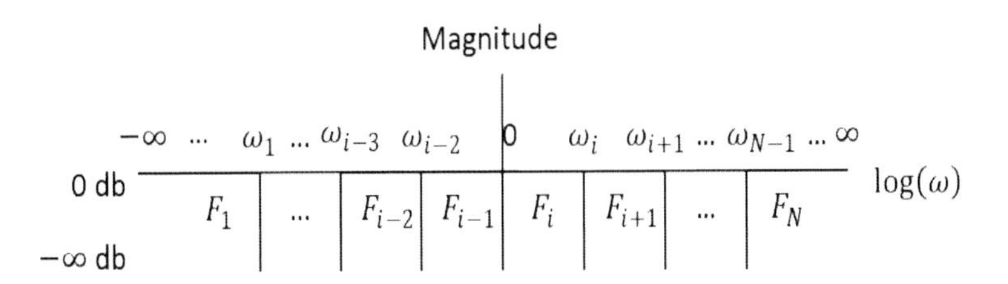


**Fig. 3.29** Ideal filter design

add to 1. The frequency 0 should be in the pass band of the first filter, and the frequency $\omega = \infty$ should be in the pass band of the last filter. Also, each filter should be equally spaced logarithmically. This design is displayed in Figure 3.29.

**Lemma 3.1.** *[22] Let $G : L_2 \to L_2$ be a linear time-invariant operator. Let $v$ be a signal in $\mathfrak{L}_{2e}$. Let $\mathbf{F} = \{F_j : j = 1, \ldots N\}$ be a set of filters. Suppose that the following conditions hold:*

- *The sum of the squares of all filters is equal to identity: $\sum_{j=1}^{N} F_j^* F_j = I$.*
- *No two filters include the same frequency: $F_j^*(j\omega)F_k(j\omega) = 0, \forall j \neq k, \forall \omega$.*

*Then,*

$$\sup_{||v||\neq 0} \frac{||Gv||^2}{||v||^2} = \max_j \sup_{||F_j v||\neq 0} \frac{||F_j Gv||^2}{||F_j v||^2} \quad .$$

The proof can be furnished using Parseval's theorem.

**Lemma 3.2.** *[22] Suppose that $F_j$, $G$, and $v$ satisfy the conditions of Lemma 3.1 and that $||F_j v||^2 \neq 0 \, \forall j = 1,...,N$. Then,*

$$\max_j \sup_{||F_j x||\neq 0} \frac{||F_j Gx||^2}{||F_j x||^2} \geq \max_j \frac{||F_j Gv||^2}{||F_j v||^2} .$$

**Lemma 3.3.** *[22] Suppose that $F_j$ is defined as in Lemma 3.1 and that $u \in \mathfrak{L}_{2e}$. Suppose that the same conditions in that lemma hold. Then,*

$$\sum_{j=1}^{N} ||F_j u||^2 = ||u||^2. \tag{3.40}$$

*Proof.* One can easily show that

$$\begin{aligned} \sum_{j=1}^{N} ||F_j u||^2 &= \sum_{j=1}^{N} \int_0^\infty u^* F_j^* F_j u \\ &= \int_0^\infty \sum_{j=1}^{N} u^* F_j^* F_j u \\ &= \int_0^\infty u^* (\sum_{j=1}^{N} F_j^* F_j) u \\ &= \int_0^\infty u^* (I) u \\ &= ||u||^2 . \end{aligned} \tag{3.41}$$

**Theorem 3.2.** *[22] Suppose that $F_j$ is defined as in Lemma 3.1 and that $x$ and $v$ are signals in $\mathfrak{L}_{2e}$. Suppose that the same conditions in that lemma hold. Then,*

$$\max_{j,||F_j v||^2} \frac{||F_j x||^2}{||F_j v||^2} \geq \frac{||x||^2}{||v||^2} \neq 0.$$

*Proof.* Assume that the maximum of the filtered gain occurs with filter $F_1$:

$$\frac{||F_1 x||^2}{||F_1 v||^2} \geq \frac{||F_j x||^2}{||F_j v||^2}, \forall j \tag{3.42}$$

and hence

$$||F_1 x||^2 ||F_1 v||^2 \geq ||F_j x||^2 ||F_j v||^2, \forall j . \tag{3.43}$$

Taking the sum over all $j$, we have

$$\sum_{j=1}^{N} ||F_1x||^2||F_jv||^2 \geq \sum_{j=1}^{N} ||F_jx||^2||F_1v||^2, \tag{3.44}$$

whence

$$||F_1x||^2 \sum_{j=1}^{N} ||F_jv||^2 \geq ||F_1v||^2 \sum_{j=1}^{N} ||F_jx||^2 . \tag{3.45}$$

Applying Lemma 3.3, we have

$$||F_1x||^2||v||^2 \geq ||F_1v||^2||x||^2 \tag{3.46}$$

or equivalently,

$$\frac{||F_1x||^2}{||F_1v||^2} \geq \frac{||x||^2}{||v||^2} . \tag{3.47}$$

**Corollary 3.1.** *Suppose that $F_j$, $G$, and $v$ are defined as in Lemma 3.1. Suppose that the same conditions in that lemma hold. Then,*

$$\max_{j,||F_jv||\neq 0} \frac{||F_jGv||^2}{||F_jv||^2} \geq \frac{||Gv||^2}{||v||^2}, ||v|| \neq 0.$$

*Proof.* This is a special case of Theorem 3.2 with $x = Gv$.

The above lemmas and a theorem provide the foundation for the main result:

**Theorem 3.3.** *[22] Suppose that $F_j$, $G$, and $v$ are defined as in Lemma 3.1. Suppose that the same conditions in that lemma hold. Then,*

$$\sup_{||v||\neq 0} \frac{||Gv||^2}{||v||^2} \geq \max_j \sup_{||F_jv||\neq 0} \frac{||F_jGv||^2}{||F_jv||^2} \geq \frac{||Gv||^2}{||v||^2}, ||v|| \neq 0.$$

*Proof.* The proof follows from Lemma 3.1, Lemma 3.2, and Corollary 3.1.

The idea of Theorem 3.3 is to improve controller falsification ability of the switching algorithm by using $\max_j \sup_{||F_jv||\neq 0} \frac{||F_jGv||^2}{||F_jv||^2}$ instead of the less tight bound $\frac{||Gv||^2}{||v||^2}$. Note that, while Theorem 3.3 may not hold for a finite time and truncated signals, it is true as the system time approaches $\infty$. For the finite time interval, [22] provides the following results.

**Theorem 3.4.** *[22] Let $G : L_2 \to L_2$ be a causal, linear, time-invariant operator, and let $v$ be a signal in $\mathfrak{L}_{2e}$. Then,*

$$||G||_\infty = \sup_{||v||\neq 0} \frac{||Gv||^2}{||v||^2} \geq \frac{||Gv||_\tau^2}{||v||_\tau^2} \; \forall v, \forall \tau, ||v_\tau||^2 \neq 0.$$

*Proof.* Since $G : L_2 \to L_2$ is a causal, linear, time-invariant operator, we have,

$$\begin{aligned}
\sup_{||v||\neq 0} \frac{||Gv||^2}{||v||^2} &\geq \sup_{||v_\tau||\neq 0} \frac{||Gv_\tau||^2}{||v_\tau||^2} \\
&\geq \sup_{||v_\tau||\neq 0} \frac{||Gv_\tau||_\tau^2}{||v_\tau||_\tau^2} \\
&= \sup_{||v_\tau||\neq 0} \frac{||Gv||_\tau^2}{||v||_\tau^2} \\
&\geq \frac{||Gv||_\tau^2}{||v||_\tau^2}, \forall ||v||_\tau \neq 0 \,. 
\end{aligned} \tag{3.48}$$

A comparison with the example by Dehghani *et al.* [26] is given next. The plant considered is an unstable LTI plant $P(s) = \frac{1}{s-1}$, while the candidate controller set is $\mathbf{C} = \{C_1 = 2, C_2 = 0.5\}$. The cost function to be used is the standard, unfiltered cost function, given by (3.3) (repeated below for convenience):

$$V(K,z,t) = \max_{\tau\in[0,t]} \frac{||u||_\tau^2 + ||\tilde{e}_K||_\tau^2}{||\tilde{v}_K||_\tau^2 + \alpha} \,.$$

In this setup, $C_1$ is a stabilizing controller, while $C_2$ is destabilizing. The Assumption 2.1 on feasibility of the control problem is clearly satisfied. Let $C_1$ be the initial controller, and let the reference signal be $r(t) = sin(t)$ for $t \geq 0$, and 0 for $t < 0$. The initial condition is $x(0) = 0$. The Hysteresis Switching Algorithm 2.1 is used, with the hysteresis constant $\varepsilon = 0.01$ and the dwell time equal to 0. The simulation was performed for $T = 50$ $s$. This setup differs from Dehghani *et al.*'s in that they examine the system in several simulations using different dwell times and initial conditions, which may be non-zero. Otherwise, it is the same. The results are shown in Figure 3.30. The top graph shows the cost $V$, while the bottom graph shows the current controller at each time.

The system is shown to take a long time (14 seconds) - to switch to the final controller, with the total of four switching times. A more ideal scenario would occur if the cost function were more sensitive in recognizing the instability of $C_2$. This can be done, for example, using bandpass filters as explained below.

A series of filters of the form:

$$F_i = \frac{1}{M_{p,i}} \frac{s}{\omega_{r,i}} \frac{\omega_{n,i}^2}{s^2 + 2\zeta\omega_{n,i}s + \omega_{n,i}^2}, i = 2, \ldots, N-1 \tag{3.49}$$

is used, where $\omega_{r,i}$ is the frequency of the resonant peak of the $i^{th}$ filter, $M_{p,i}$ is the maximum value of the peak, $\zeta = 0.2$, and $\omega_{n,i} = \frac{\omega_{r,i}}{\sqrt{(1-2\zeta^2)}}$. The $\omega_{r,i}$ are chosen to be 20 logarithmically equally spaced values between 1 Hz and 100 Hz. The simulation was performed next, using the same setup as the one in Figure 3.30, and the results are shown in Figure 3.31.

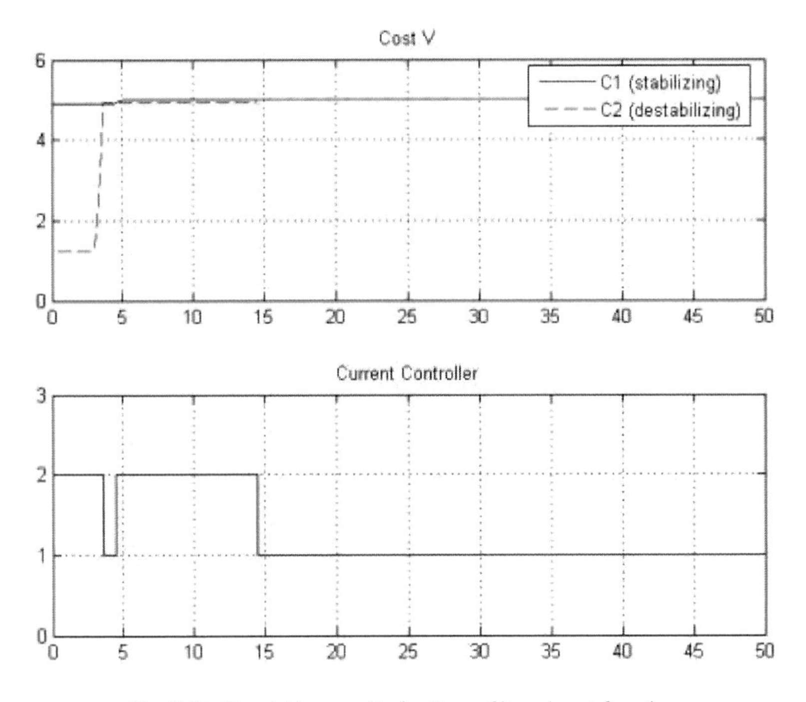


**Fig. 3.30** Simulation results for the unfiltered cost function

It can be seen that the system with the first order filters took less time to settle (4 seconds) than the system without filters. However, some chattering takes place. Increasing the order of the filters, as in:

$$F_i = \frac{1}{M_{p,i}} \frac{2s^2}{(s+\omega_{r,i})^2} \frac{\omega_{n,i}^2}{s^2 + 2\zeta\omega_{n,i}s + \omega_{n,i}^2}, i = 2,...,N-1 \tag{3.50}$$

with $F_1 = \frac{1}{(s+1)^2}$ and $F_N = \frac{s^2}{(s+100)^2}$, removes the chattering effect (see Figure 3.32), while slightly increasing the settling time. The hypothesis made in [22] is that the use of higher order filters resulted in increasing the cost function faster and higher, as the signals resonated with a particular filter's pass band frequency. Thus, the destabilizing or non-optimal controllers are falsified sooner.

It should be noted that using filtered cost functions can be beneficial over a limited frequency range (reducing settling time is the most conspicuous benefit). A high frequency filter would result a prohibitively small sampling time, while using a low frequency filter would result in a longer simulation time. Another property of interest, such as reducing the total number of switches, could perhaps be addressed through the use of different type of filters, such as Chebyshev.

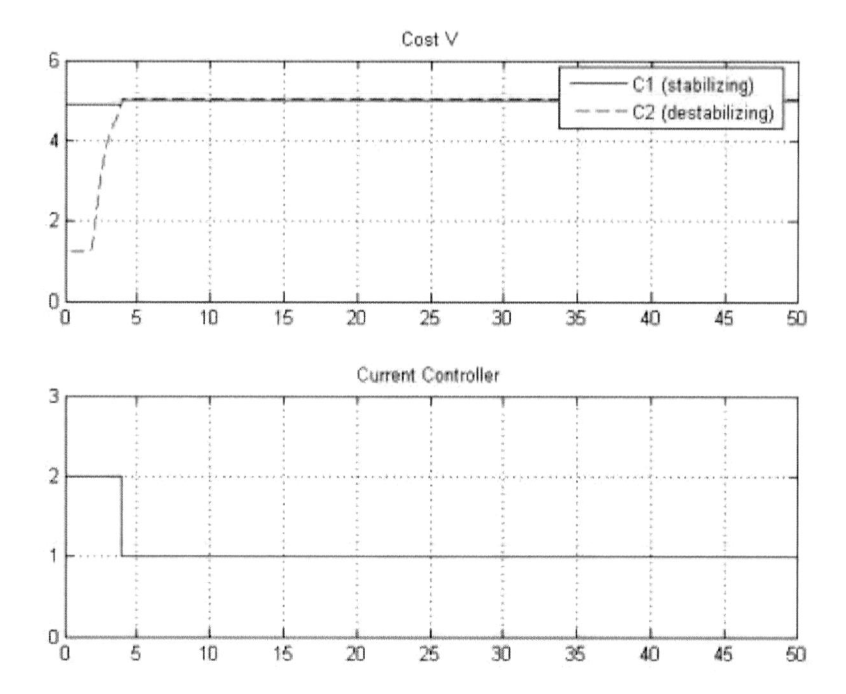


**Fig. 3.31** Simulation results with first order filters

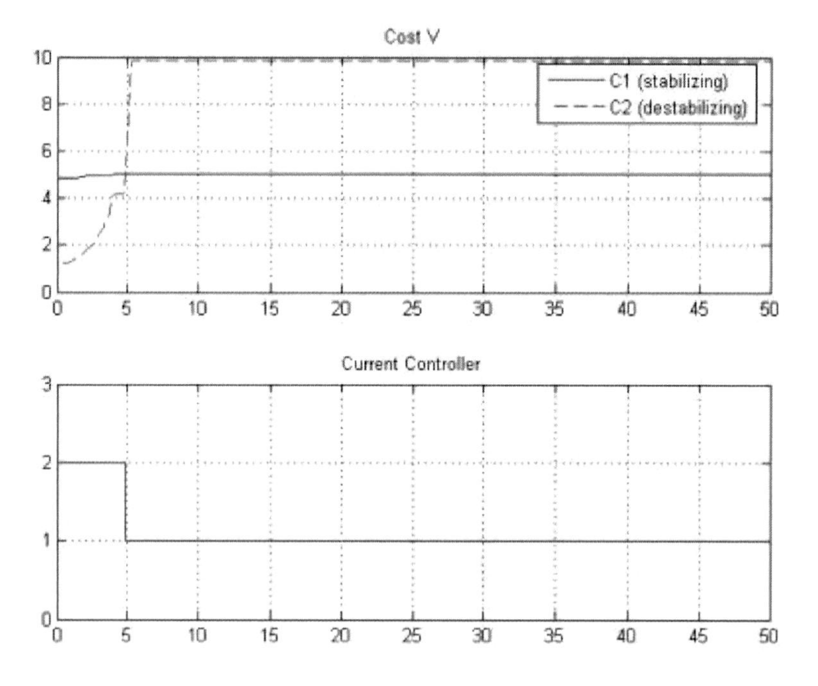


**Fig. 3.32** Simulation Results With Second Order Filters

## 3.4 Applications of Unfalsified Control

Various applications of the unfalsified concept have arisen in the last decade, both by the authors of this monograph and their collaborators, and independently by other researchers. To give insight into these various observed aspects of the theory, we provide some of the representative results.

### 3.4.1 *Automatic PID Tuning*

In [54], a method based on the early version of the unfalsified control concept was designed to adaptively tune PID parameters. With the improper derivative term approximated as in

$$u = (k_P + \frac{k_I}{s})(r - y) - \frac{sk_D}{\varepsilon s + 1} \tag{3.51}$$

and with the control gains $k_P > 0$, $k_I, k_D \geq 0$, the unfalsification PID controller scheme can be represented as shown in Figure 3.33:

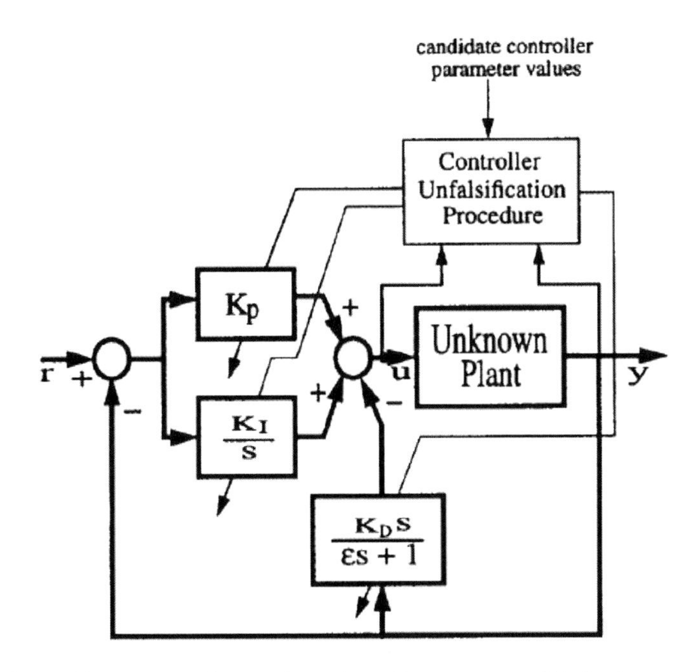


**Fig. 3.33** PID controller configuration with approximated derivative term

Standard PID controllers have causally left invertible (CLI) property, hence there is a unique fictitious reference signal $\tilde{r}_i(t)$ associated with each candidate controllers, computed in real time by filtering the measurement data $(u, y)$ via the expression:

$$\tilde{r}_i = y + \frac{s}{sk_{P_i} + k_{I_i}} \left( u + \frac{sk_{D_i}}{\varepsilon s+1} y \right) . \tag{3.52}$$

According to the definition of the fictitious reference signal, only measured output data are used for its computation (Figure 3.34):

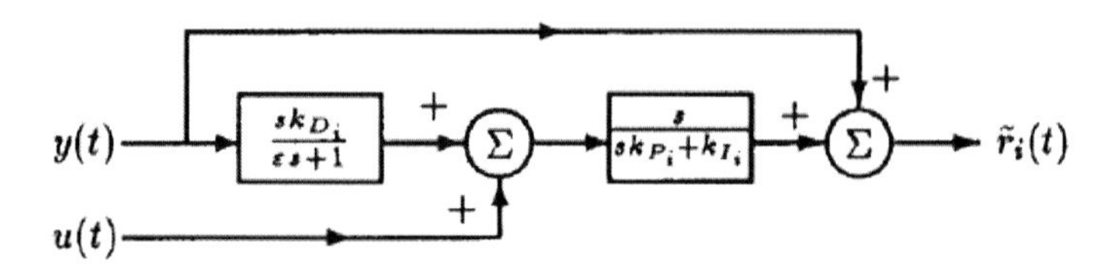


**Fig. 3.34** Generating the $i^{th}$ fictitious reference signal $\tilde{r}_i(t)$

The candidate controller set $K$ containing the parameters $k_P, k_I$, and $k_D$ is chosen as a finite set of parameter triples, and the performance specification (cost function) is an integral performance inequality of the form:

$$\tilde{J}_i(t) = -\rho + \int_0^t \Gamma_{spec}(\tilde{r}_i(t), y(t), u(t))dt, \ \ \forall t \in \left[0, \tau\right] \tag{3.53}$$

where $u(t), y(t)$, and $t \in \left[0, \tau\right]$ are the measured past plant data. A discretized equivalent of the cost function is:

$$\begin{aligned} \tilde{J}_i(k\Delta t) &= \tilde{J}_i((k-1)\Delta t) + \int_{(k-1)\Delta t}^{k\Delta t} \Gamma_{spec}(\tilde{r}_i(t), y(t), u(t))dt \\ &\approx \tilde{J}_i((k-1)\Delta t) + \frac{1}{2}\Delta t\{\Gamma_{spec}(\tilde{r}_i(k\Delta t), y(k\Delta t), u(k\Delta t)) + \\ &\quad + \Gamma_{spec}(\tilde{r}_i((k-1)\Delta t), y((k-1)\Delta t), u((k-1)\Delta t))\} \end{aligned} \tag{3.54}$$

when $\rho = 0$.

The simulation performed in [54] makes use of the following performance specification:

$$\Gamma_{spec}(\tilde{r}_i(t), y(t), u(t)) = \|\omega_1 * (\tilde{r}(t) - y(t))\|^2 + \|\omega_2 * u(t)\|^2 - \sigma^2 - \|\tilde{r}(t)\|^2 \tag{3.55}$$

where $\sigma$ is a constant representing the root-mean-square effects of noise on the cost, and $\omega_1$, and $\omega_2$ are the weighting filters.

Notice that, in this early study on unfalsified control, the crucial cost detectability property was not yet recognized. It is important to note, however, that the example in this study demonstrated an efficient manner in which to prune "bad" controllers and reach the "good" one, if such controller is at the designer's disposal.

The controller unfalsification procedure determines, at each sample instant $\tau = k\Delta t$, which previously unfalsified controllers are now falsified based on the consistency test:

$$\tilde{J}_i(k\Delta t) \leq 0$$

If at a particular time instant a switching to a new controller occurs, then the algorithm in [54] resets the states of the controller, thus preventing any discontinuity in the resulting output signals. In this way, adverse overshoot in the signals $u, y$ is avoided.

The (unknown) plant considered in this simulation is $P(s) = \frac{2s^2+2s+10}{(s-1)(s^2+2s+100)}$. The weighting filters are chosen as $W_1(s) = \frac{s+10}{2(s+0.1)}$ and $W_2(s) = \frac{0.01}{1.2(s+1)^3}$. As a reference input, a step is chosen $r(t) = 1, \forall t \geq 0$. Without loss of generality, all initial conditions were set to zero. The sampling time is $\Delta t = 0.05$ $s$; the small positive constant $\varepsilon$ is set to 0.01; no noise is considered ($\sigma = 0$), and the candidate set of the controller parameters is the union of the sets $K_D = \{0.6, 0.5\}$, $K_P = \{5, 10, 25, 80, 110\}$ and $K_I = \{2, 50, 100\}$, making the overall number of distinct PID controllers equal to 30.

The simulation of the above described unfalsified switching adaptive control algorithm yields the results shown in Figures 3.35, 3.36 and 3.37 below.

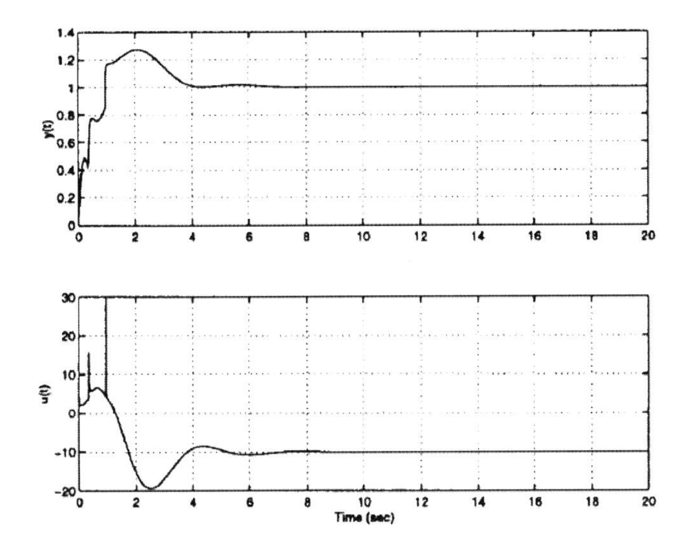


**Fig. 3.35** Plots of signals $y(t)$ and $u(t)$ when the states of the controller are not reset at the switching time. Poor transients with spikes can occur if controller states are not properly reset at switching time.

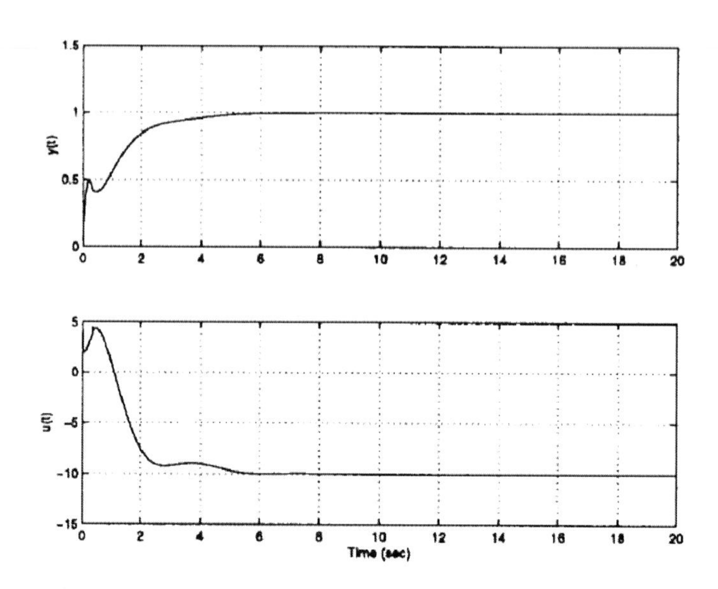


**Fig. 3.36** Simulation results showing good transient response with correctly reset controller states

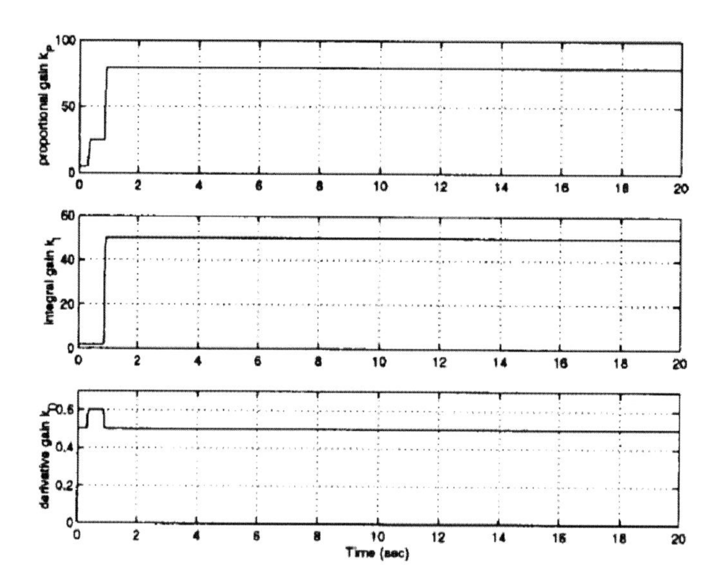


**Fig. 3.37** Simulation results showing the changes in controller gains

### 3.4.2 Unfalsified Adaptive Spacecraft Attitude Control

The problem of antenna mapping control using unfalsified control theory was considered in [100]. The problem is particularly amenable to the application of the data-driven unfalsified theory due to a plethora of uncertainties in the spacecraft dynamics. Instead of making attempts to provide accurate plant models that will drive the selection of the controller, it is precisely this uncertainty in the plant, which is used to falsify the currently active controller and switch to an alternative one from the set of available controllers.

The plant is described in the following way. High-powered communication satellites, such as the Boeing Satellite Systems 702 spacecraft illustrated in Figure 3.38, are characterized by flexible appendages whose modal frequencies are uncertain and whose modal masses are often significant. The presence of this large uncertainty in the spacecraft dynamics motivates the control system design presented in this section. The flexible appendage modes are characterized by 180° phase shifts and typically a significant gain, whereas the modal frequency uncertainty may be 25% or greater. Familiar techniques using simple roll-off gain stabilization filters or notch filter phase stabilization would normally provide robust stabilization where the modal frequencies either decades higher than the desired attitude control bandwidths or simply less uncertain, for notch filter stabilization. However, these problems cannot be neglected in the spacecraft operation, where the the attitude control laws typically require a relatively fast closed loop response. The peculiarity of this problem stems from the fact that 0.05 Hz, which in many applications would be considered extremely low, may be considered "high" bandwidth in this context where the first mode frequency may be 0.1 Hz or lower and other physical limitations such as actuator capabilities may limit bandwidth in their linear range. In addition, a common source of excitation of the flexible modes in one of those high bandwidth modes stems from the "on-off" characteristic of the chemical propulsion Reaction Control System (RCS) thrusters used as primary actuators.

One such high bandwidth mode of operation for geosynchronous communication satellites is antenna mapping. This procedure occurs during spacecraft initialization after transfer orbit operations have successfully deposited the spacecraft into its final orbital slot. The procedure generally consists of a series of slews whose purpose is to scan the payload antenna pattern across receivers pre-positioned on the ground to measure received power. Each slew provides a cross-section of the antenna gain pattern (the cross-section often referred to as a "cut"). Stable, high bandwidth attitude control to accomplish the slews, coupled with precise knowledge of spacecraft attitude obtained from Inertial Reference Unit (IRU) measurements (appropriately calibrated before the slews commence) provides all the necessary information to infer the shape of the antenna pattern along with its orientation in a spacecraft-specific coordinate frame (also referred to as the body frame). The appropriate attitude command biases can subsequently be applied to ensure the proper antenna pattern coverage on the ground.

In the simplified simulation model for the antenna mapping control problem, there are two channels considered, the azimuth and elevation channels. The flexible

**Fig. 3.38** Boeing 702 6-panel spacecraft

mode is assumed to appear only in the elevation channel with one frequency and zero damping ratio. The quantization and measurement noise are assumed to be present for the position measurements. The proportional derivative (PD) control law used is given as:

$$u = C_p(r - y) - C_d\dot{y} \tag{3.56}$$

and the performance criterion (cost function) is chosen as:

$$\begin{aligned} \|T_{er}(j\omega)\| &< \|W_1^{-1}(j\omega)\| \\ \|T_{ur}(j\omega)\| &< \|W_2^{-1}(j\omega)\| \end{aligned} \tag{3.57}$$

where $T_{er}$ is the transfer function from $r$ to $e = r - y$, and $T_{ur}$ is the transfer function from $r$ to $u$, whereas $W_1$ and $W_2$ are weighting filters specifying the error and control responses in frequency domain. The time domain equivalent description of this criterion reads as:

$$\begin{aligned} \|\hat{w}_1 * (r-y)\|_t &< \|r\|_t \\ \|\hat{w}_2 * u\|_t &< \|r\|_t, \quad \forall t, \forall r \end{aligned} \tag{3.58}$$

where $\hat{w}_i$ is the impulse response of the filter $W_i$, $i \in [1,2]$.

From (3.56), the reference signal can be back-computed as:

$$r = \frac{1}{C_p}u + \frac{C_d}{C_p}\dot{y} + y \equiv \theta^T \psi + y \tag{3.59}$$

where

$$\theta = \left[\tfrac{1}{C_p},\ \tfrac{C_d}{C_p}\right]^T$$

is the controller parameter vector and $\psi = \left[u,\ \dot{y}\right]^T$ consists of the plant input/output measurements. For a particular candidate controller parameter vector $\theta$, Equation 3.59 produces a fictitious reference signal as per Definition 2.6, denoted $\tilde{r}$. Hence, falsification of this controller from a candidate set occurs when (3.58) fails to hold at any time $\tau$ for the computed $\tilde{r}$. Conversely, if (3.58) holds for all times up to present time, then the controller parameterized by that particular $\theta$ is unfalsified by measurements.

Rearrangement of the performance criterion (3.58) and the fictitious signal (3.59) equations leads to the following quadratic inequalities:

$$\theta^T A_t^e \theta - 2\theta^T B_t^e + C_t^e < 0 \tag{3.60}$$

$$\theta^T A_t^u \theta - 2\theta^T B_t^u + C_t^u < 0 \tag{3.61}$$

where

$$\begin{aligned} A_t^e &= \int_0^t \xi\xi^T - \psi\psi^T d\tau; \ B_t^e = -\int_0^t \psi y d\tau \\ C_t^e &= \int_0^t y^2 d\tau; \ A_t^u = -\int_0^t \psi\psi^T d\tau \\ B_t^u &= -\int_0^t \psi y d\tau; \ C_t^u = \int_0^t (\hat{w}_2 * u)^2 - y^2 d\tau \\ \xi &= \hat{w}_1 * \psi\,. \end{aligned} \tag{3.62}$$

At any time $t$, the set of unfalsified controller parameter vectors is obtained as the intersection

$$\bigcap_{0<\tau<t} (K_\tau^e(\theta) \cap K_\tau^u(\theta))$$

where $K_\tau^x(\theta) = \left\{\theta | \theta^T A_\tau^x \theta - 2\theta^T B_\tau^x + C_\tau^x < 0\right\}$. At any given time, the intersection of the above inequalities yields an as-yet-unfalsified section in the parameter space as shown in Figure 3.39.

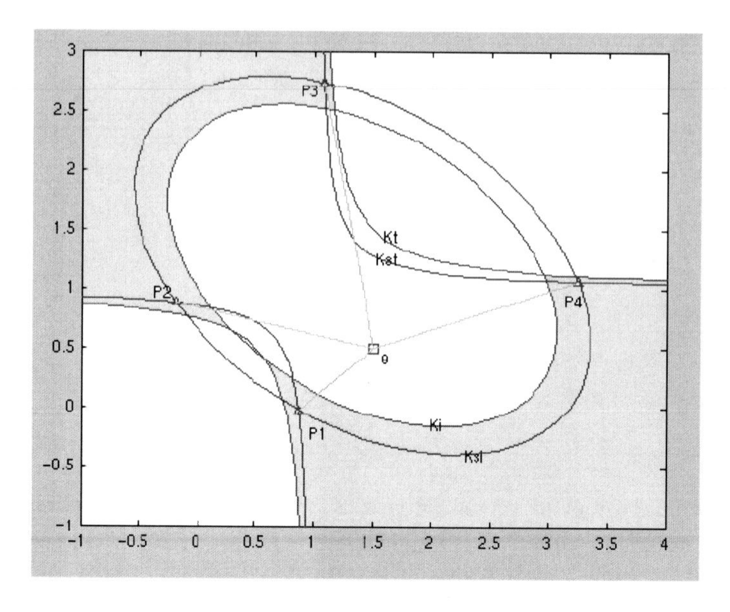


**Fig. 3.39** The yellow area represents the region of unfalsified control parameter vectors

For the simulation, the following constraints on the controller parameter space are used in [100]: $\theta_1 \in [5e-6, 1e4]$, $\theta_2 \in [0, 1e4]$. Figure 3.40 shows the time history of the switching in the controller parameters. The last switch in this experiment occurs around $t = 800\ s$. The set of currently unfalsified controllers is found at each time instant as the intersection of quadratic inequalities, as shown in Figure 3.41.

In Figure 3.42, the magnitude of the achieved transfer function $T_{er}$ lies below that of the inverse of the weighting transfer function $W_1$ used in the unfalsified algorithm. Also shown is the error transfer function magnitude of the non-adaptive nominal proportional-integral (PI) controller parameters used for comparison. Similar observations can be drawn about the magnitude of the achieved transfer function $T_{ur}$ in Figure 3.43.

Figures 3.44 and 3.45 show the azimuth channel reference signal and the azimuth pointing error from the unfalsified algorithm. Figure 3.46 is the three wheel control signal.

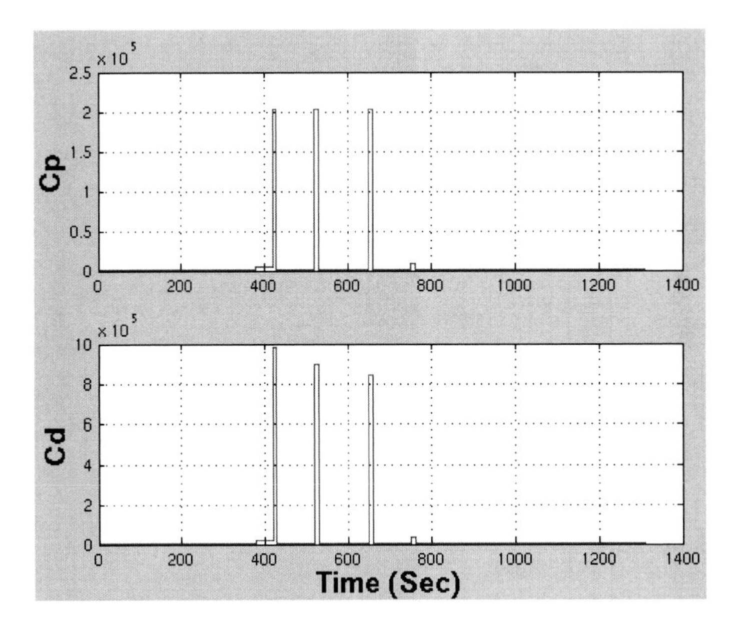


**Fig. 3.40** The last switching of unfalsified controller parameters in azimuth channel is about 800 $s$

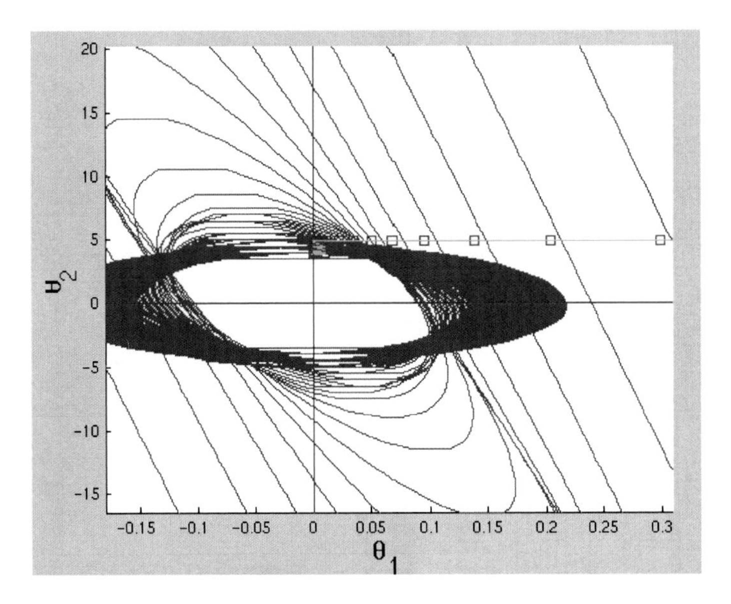


**Fig. 3.41** Switching of the controller parameter vector

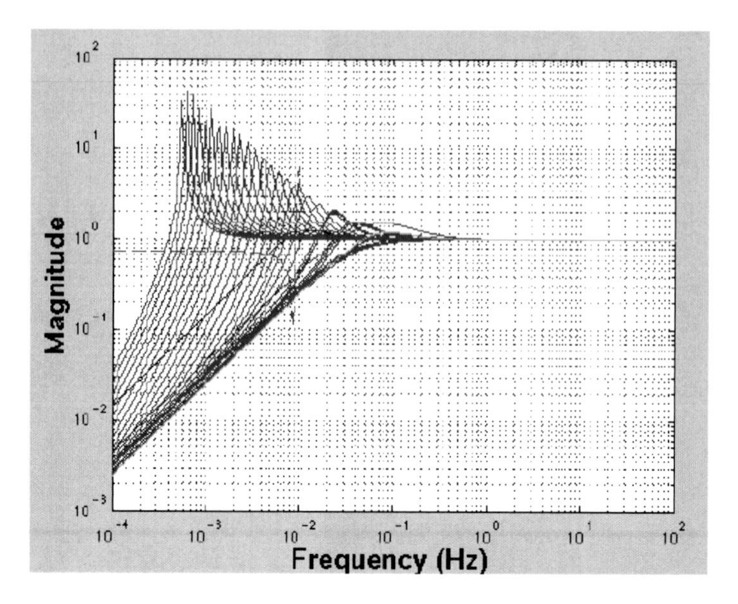


**Fig. 3.42** The magnitude of the achieved closed loop error transfer function $T_{er}$ (green) lies below that of $W_1^{-1}$(red). The error transfer function magnitude for nominal non-adaptive parameters used for comparison is shown in black. The initial error transfer function magnitude is shown by the broken blue line.

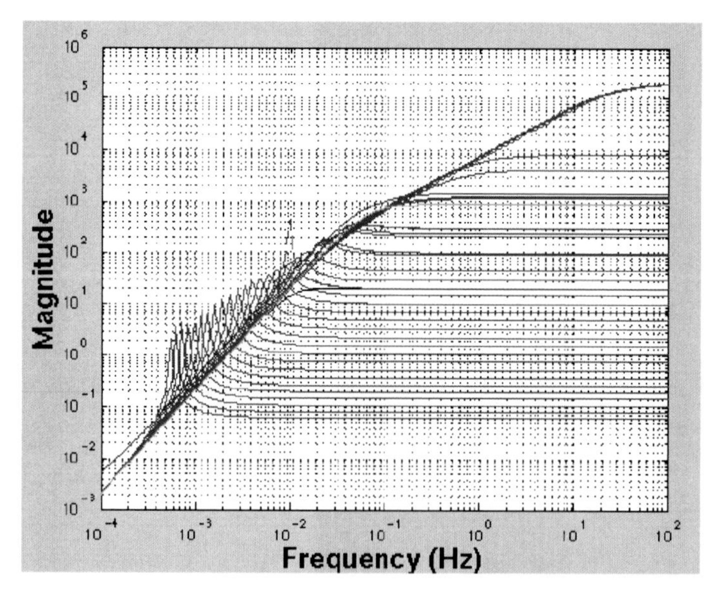


**Fig. 3.43** The magnitude of the achieved closed loop control transfer function $T_{ur}$ (green) lies below that of the inverse of the weighting transfer function $W_2$ (red). The control transfer function magnitude for nominal non-adaptive parameters used for comparison is shown in black. The initial error transfer function magnitude is shown by the broken blue line.

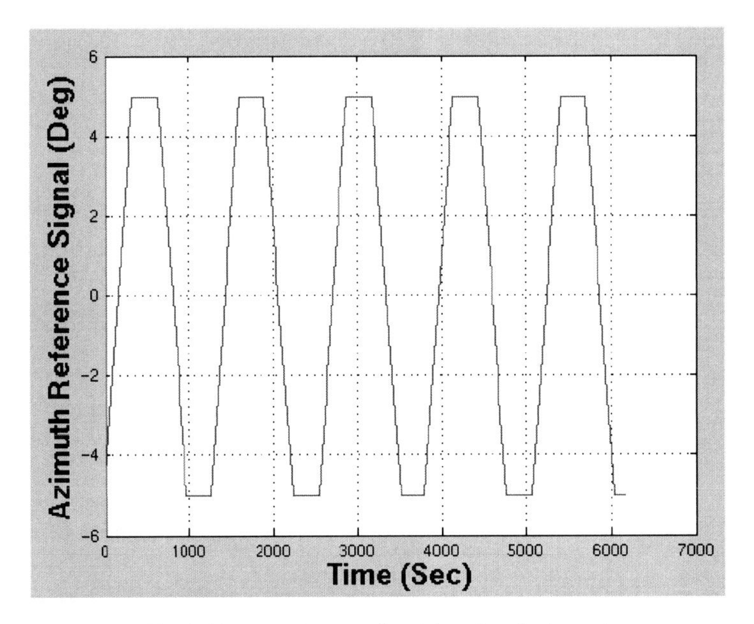


**Fig. 3.44** The reference signal in azimuth channel

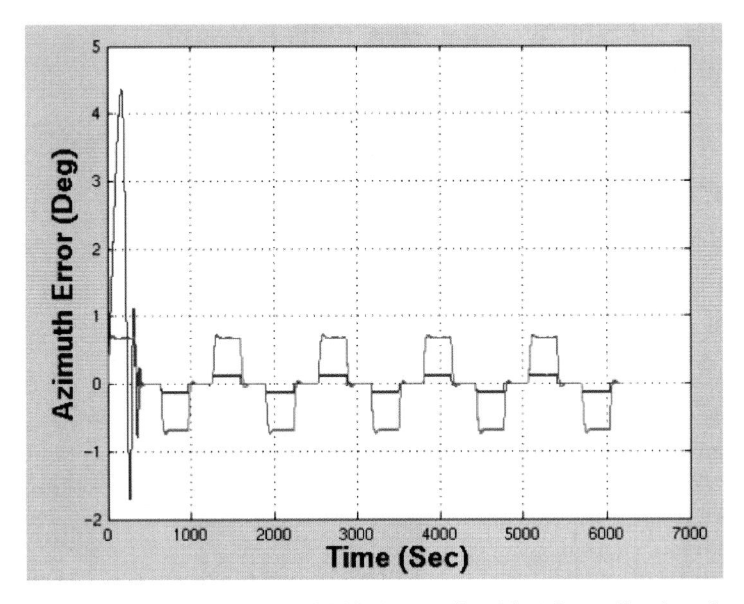


**Fig. 3.45** The azimuth error of the unfalsified controller (blue) is smaller than that of the nominal non-adaptive controller (green)

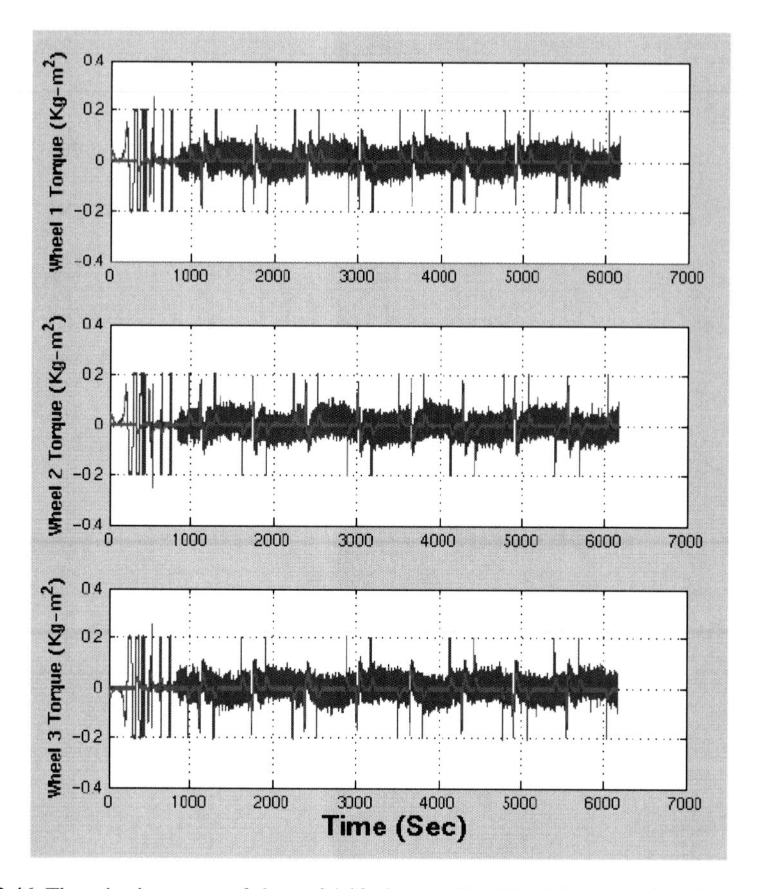


**Fig. 3.46** The wheel torques of the unfalsified controller (blue) is larger than those of the nominal non-adaptive controller (green)

### *3.4.3 Switching Control of Missile Autopilot*

In another example of applicability of the safe switching adaptive control paradigm, we relate the details of [16], a study of the robust missile autopilot control design. Owing to the highly nonlinear dynamics, wide variations in plant parameters, and strict performance requirements related to the highly maneuverable missiles, the missile autopilot design is known to pose significant challenges.

Specifically, the control objective in [16] is to design a longitudinal autopilot for a tail-governed missile, *i.e.*, use the tail deflection to track an acceleration maneuver with a time constant of less than 0.35 $s$, a steady state error of less than 5 %, and a maximum overshot of 20 % for the step response. The autopilot is supposed to provide such performance over the operation range of $+/-0.35$ $rad$ angle of attack.

The candidate controller set has been chosen so as to have enough structure to guarantee a simple falsification procedure, and at the same time, include as a special case the PID controller, which is commonly used in the industry because of its simplicity and good performance. The Figure 3.47 shows a typical PID missile autopilot structure.

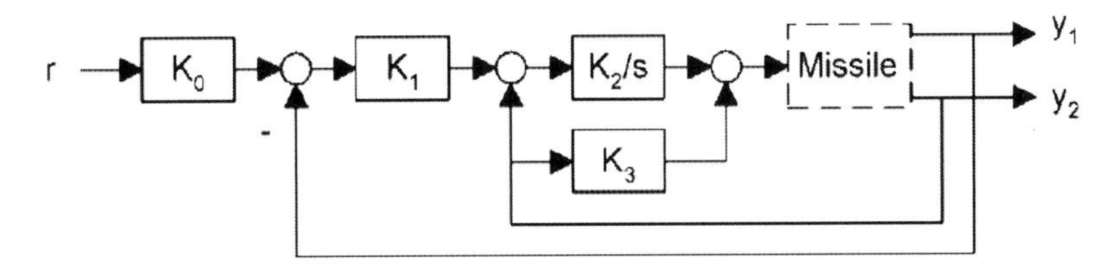


**Fig. 3.47** PID missile autopilot

The controller structure for the unfalsified control that includes the PID as a special case is shown in Figure 3.48. where $\frac{N_1(s)}{D_1(s)}$, and $\frac{N_2(s)}{D_2(s)}$ are stable minimum phase bi-proper transfer functions, and $H$ is a pure gain.

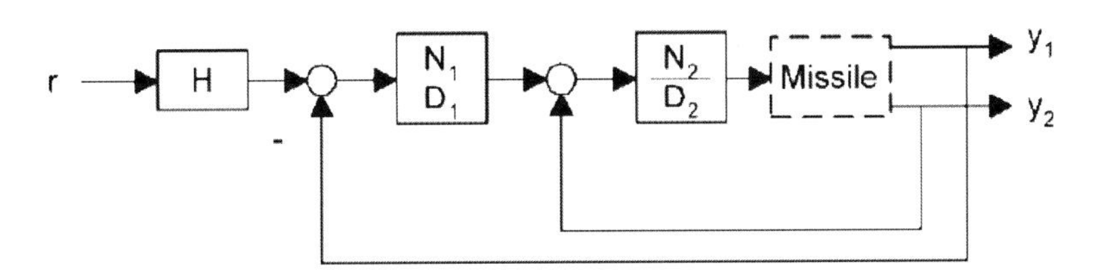


**Fig. 3.48** Controller structure

The set of candidate controllers is generated by allowing the coefficients of $D_1(s)$ and $D_2(s)$ and the gain $H$ to vary in some intervals around some nominal values. In order to reduce the number of computation, each coefficient takes a finite number of values in its interval. The nominal values are calculated beforehand by a classical control technique based on simplified model. Thus, the set of candidate controller can be represented as:

$$\mathbf{K} = \{(r,u,y) \in \mathbb{R} \times U \times Y | u = \frac{N_2(s)}{D_2^{\theta}(s)} [\frac{N_1(s)}{D_1^{\theta}(s)} (H^{\theta} r - y_1) + y_2], \theta \in \Theta\} .$$

The performance specification set in [16] has the form:

$$T_{spec} = \{(r,y,u) | J(r,u,y,\tau) \geq 0, \ \forall \tau \geq 0\}$$

where the cost function $J(r,y,u,\tau)$ is chosen as follows:

$$J(r,y,u,\tau) = [-|r - y_1| + |E_{ss} r| + |\frac{Ks}{s+N} r|]|_{t=\tau} .$$

This cost is defined to shape the time response such that it defines upper and lower bounds for the tracking error. The value of the bound depends on the reference signal that is being tracked. In particular, the bounds are the sum of the $E_{ss}$ % of the reference signal and a term proportional to the derivative of the reference signal. The first term signifies the steady state requirement ($E_{ss}$ stands for steady state error), and the second one signifies the transient response requirement. The second term can be written as:

$$\frac{Ks}{s+N}r = K(1 - \frac{1}{\frac{s}{N}+1})r$$

which is proportional to the error of a first order system. Moreover, this cost can be interpreted as the model of the worst closed loop performance acceptable, such that the controller has to provide a performance that is at least as good as this model.

A nonlinear mathematical model described in the existing literature is used. In particular, a pitch axis model of a missile is used, which flies at Mach 3 and at an altitude of $20,000\ ft$. This mathematical model is described in [84], and owing to its realistic representation of the dynamics, it has been extensively used as a benchmark for missile analysis and controller design. The state equations of this model are:

$$\dot{\alpha} = \cos(\alpha)K_{\alpha}MC_n(\alpha,\delta,M) + q$$
$$\dot{q} = K_q M^2 C_m(\alpha,\delta,M)$$

and the output equation is:

$$\eta = \frac{K_z}{g}M^2 C_n(\alpha,\delta,M)$$

where $\eta$ is the acceleration in $g$, $\delta$ is the tail deflection in $rad$, $\alpha$ is the angle of attack in $rad$, $q$ is the pitch rate in the plane $(G_x, G_z)$ in $rad/s$, $M$ is Mach number, and the stability derivatives are:

$$C_n(\alpha,\delta,M) = a_n\alpha^3 + b_n|\alpha|\alpha + c_n(2 - \frac{M}{3})\alpha + d_n\delta,$$
$$C_m(\alpha,\delta,M) = a_n\alpha^3 + b_m|\alpha|\alpha + c_m(-7 + \frac{8M}{3})\alpha + d_m\delta\ .$$

The actuator is modeled as a second order linear transfer function:

$$A(s) = \frac{\omega_a^2}{s^2 + 2\zeta_a\omega_a s + \omega_a^2}.$$

For the simulation, the parameters take the values in Figure 3.49, which correspond to a missile flying at Mach 3 at $20,000\ ft$.

$K_a = 0.7 P_0 S / m / V$ $\quad S = 0.44$ $\quad a_n = 1.0286 * 10^{-4}$

$K_q = 0.7 P_0 S d / I_y$ $\quad m = 13.98$ $\quad b_n = -0.94457 * 10^{-2}$

$K_z = 0.7 P_0 S / m$ $\quad V = 1036.4$ $\quad c_n = -0.1696$

$P_0 = 973.3$ $\quad d = 0.75$ $\quad d_n = -0.034$

$a_m = 2.1524 * 10^{-4}$ $\quad g = 32.2$

$b_m = -1.9546 * 10^{-2}$ $\quad \varpi_a = 150$

$c_m = 0.051$ $\quad \xi_a = 0.7$

$d_m = -0.206$

**Fig. 3.49** Simulation parameter values

The controller is parameterized as follows:

$$u = \frac{N_2(s)}{D_2^{\theta}(s)} \left[ \frac{N_1(s)}{D_1^{\theta}(s)} (H^{\theta} r - y_1) + y_2 \right]$$

where the polynomials involved are chosen as follows:

$$N_1(s) = \frac{s}{25} + 1$$

$$N_2(s) = \frac{s}{5} + 1$$

$$D_1^{\theta}(s) = \frac{\theta_1 s + \theta_2}{\theta_5}$$

$$D_2^{\theta}(s) = \theta_3 s + \theta_4$$

$$H^{\theta} = \frac{1}{\theta_5} .$$

In this case, the candidate controller set is generated by allowing the parameters to vary up to 20 % around a nominal value, which is calculated by a classical technique based on a linearized missile model. In particular, each controller takes discrete values in the intervals. For simplicity of simulation, five values for each parameter are considered which gives 3125 controllers. The nominal controller used in the simulations has the following values:

$$\theta_1 = 0.4,\ \theta_2 = 0.01,\ \theta_3 = 10.2,\ \theta_4 = 2,\ \theta_5 = 0.9.$$

Recall that the performance specification is given by:

$$\left|\frac{Ks}{s+N} r\right| + |E_{ss} r| - |r - y_1| \geq 0 .$$

From the missile steady state error requirement, $E_{ss}$ is selected to be 0.05 (5 %), and from the transient response requirements on overshoot and time constants $K$ and $N$ are chosen as 1.2 and 4 respectively. This selection allows some non-minimum phase type behavior. According to the switching rule described in [16], when the existing controller is falsified, it is replaced with the controller that has the smallest average error.

Nonlinear simulation results of the unfalsified control autopilot are shown next. Figure 3.50 shows the reference signal, the output signal, and the bounds on the response; Figure 3.51 represents the control signal; Figure 3.52 shows the evolution of the controller parameter set, and finally, Figure 3.53 shows the evolution of the number of controllers.

In Figure 3.50, it can be seen that during most of the time the output signal stays inside the bounds, but at some points it goes outside. At these points, the current controller is falsified and replaced with the one having the smallest average tracking error from the set of as-yet-unfalsified candidate controllers.

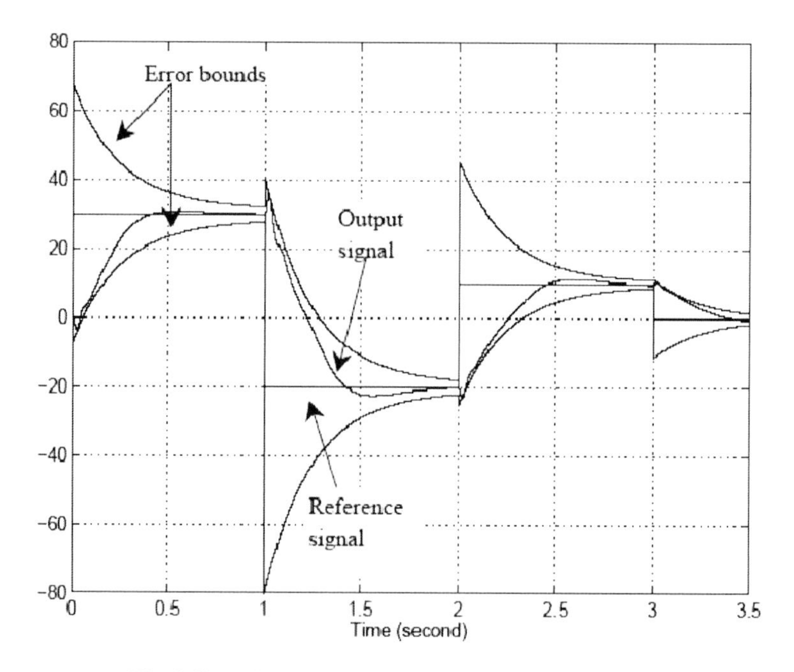


**Fig. 3.50** Reference signal, output signal, and error bounds

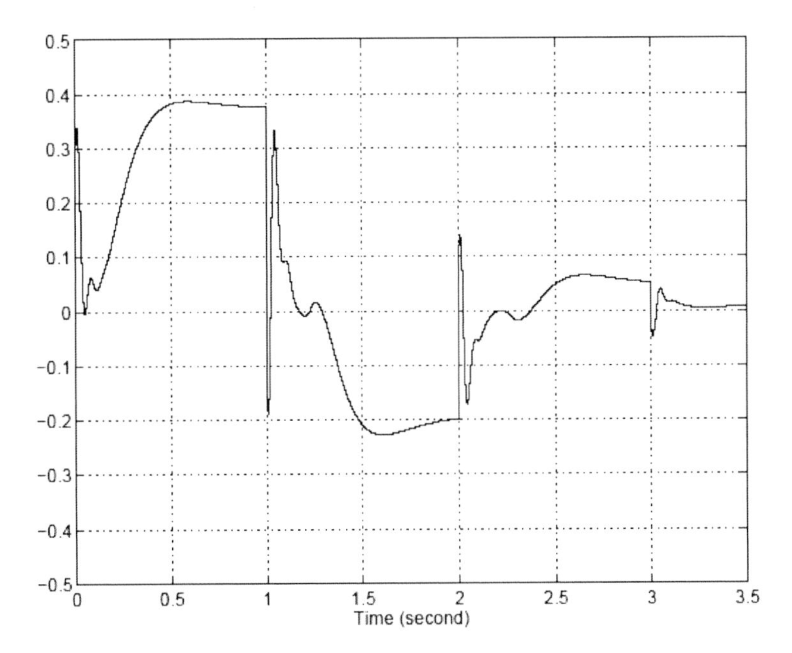


**Fig. 3.51** Control signal

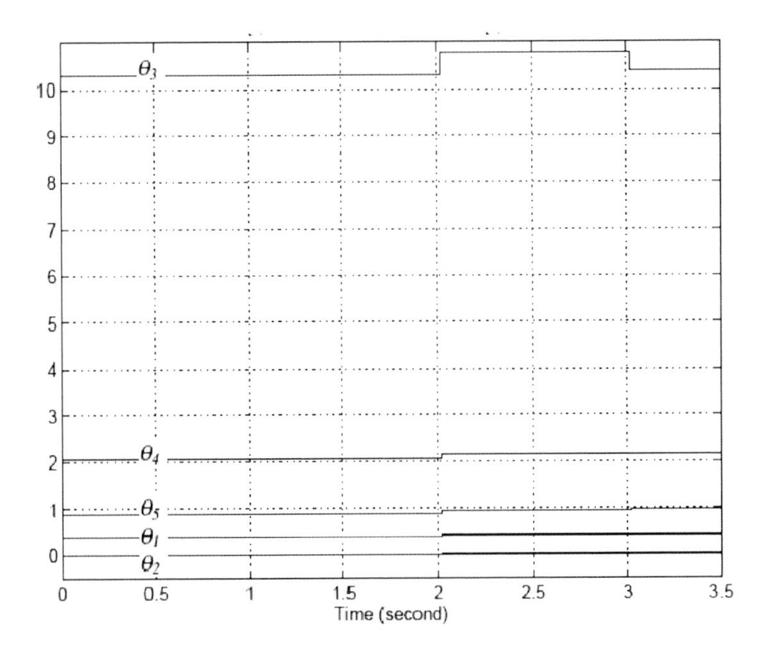


**Fig. 3.52** Evolution of Controller parameters

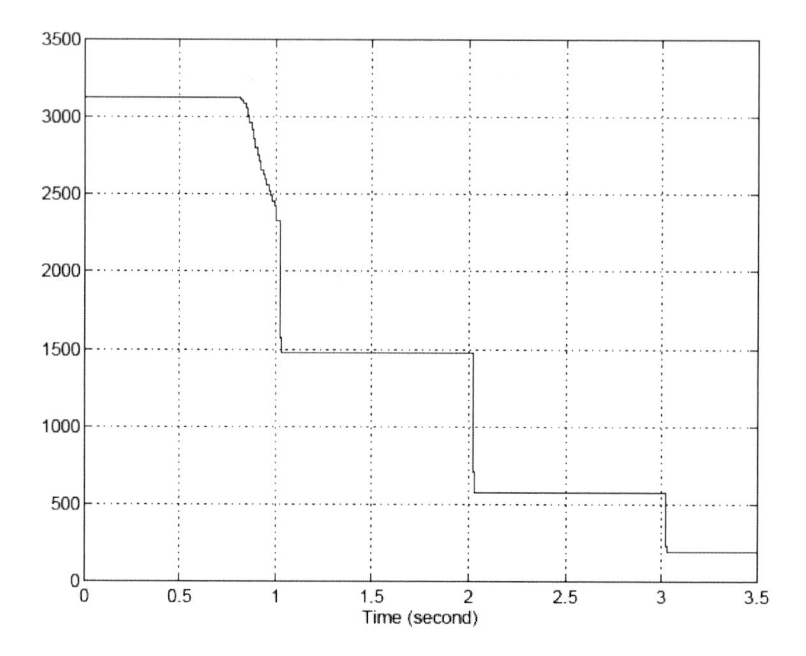


**Fig. 3.53** Evolution of the number of unfalsified controllers

In Figure 3.53, one can see the evolution of the unfalsified number of controllers. This figure also shows that there are 198 controllers that have not been falsified, which means they could have met the performance if put in the loop.

### *3.4.4 Switching Congestion Control for Satellite TCP/AQM Networks*

In [21], a congestion controller using data-driven, safe switching control theory is used to improve the dynamic performance of the satellite TCP/AQM (Transmission Control Protocol/Active Queue Management) networks. A Proportional-Integral-Derivative (PID) controller whose parameters are adaptively tuned by switching among members of a given candidate set using observed plant data is compared with some classical AQM policy examples, such as the Random Early Detection (RED) and fixed Proportional-Integral (PI) control. A new cost detectable switching law with the interval cost function switching algorithm, which improves the performance and also saves the computational cost, is developed and compared with a law commonly used in the switching control literature. Finite-gain stability of the system is proved. The Smith predictor and the TCP NewReno algorithm are also incorporated to further improve the performance. Simulations are presented to validate the theory.

To lay out the foundation of the problem, we note that satellite networks play an important role in broadcasting data over large geographic locations, and are an effective means for reaching remote locations lacking in communication infrastructure. Both the military and civilian applications benefit immensely from the use of satellite networks. For example, national defense depends on satellite communications for robust, rapidly deployable and secure communications in hostile environments. Among the civilian applications, supplying rural locations with high data rate communication services is currently feasible only through affordable satellite communication, because of the lack of availability in fiber networks.

Congestion on the satellite networks is one of the major communication problems. TCP congestion control algorithms use packet-loss and packet-delay measurements respectively to detect congestion [98]. Recently, Active Queue Management (AQM) has been proposed to support the end-to-end congestion control in the Internet, by sensing impending congestion before it occurs and providing feedback information to senders by either dropping or marking packets, so that congestion, causing a significant degradation in network performance, can be avoided [1]. In the control theory, AQM can be considered as a nonlinear feedback control system with a delay.

The study of congestion problem within time delay systems framework has been successfully exploited using control theory. In [1], [48], [63], dynamical models of the average TCP window size and the queue size in the bottleneck router are derived, and linearized at some equilibrium point, and the PI congestion controllers are designed. In [56], a delay dependent state feedback controller is proposed by means of compensation of the delay with a memory feedback control. The latter methodology is interesting in theory but there are indications that it is not very suitable in practice. In [64], robust AQMs are derived using a time delay system approach. Nevertheless, in the above cited study, only a single specific model was considered, which means that the designed controller is either effective only near the nominal point, or conservative, which results in sluggish responsiveness.

In [21], a new AQM algorithm for long fat networks and in particular multilayer satellite networks is presented, and the control theory on congestion problem is extended by applying unfalsified Safe Switching Control(SSC) theory. The main goal is to illustrate the potential impact of the SSC methodology on TCP/AQM networks with dynamically varying parameters and time delays, as is the case in long fat satellite networks. The scheme has two features: multiple controllers and adaptive tuning to the optimal controller. Through Matlab and ns-2 simulations, comparison of the performance of SSC with that of Random Early Detection (RED) [34], adaptive RED [32], [33], and fixed PI controller [48] is performed. From the simulation results, it can be shown that the SSC scheme can adaptively deal with the change of the number of connections and heterogeneous delays.

To design congestion controllers, large scale networks are often simplified according to a well-known dumbbell topology, shown in Figure 3.54. The network consists of $n$ senders, one bottleneck router and one receiver, which in a cluster form of satellite networks correspond to normal satellite nodes, cluster-head of one cluster and a cluster-head of another (adjacent) cluster, respectively. As we are

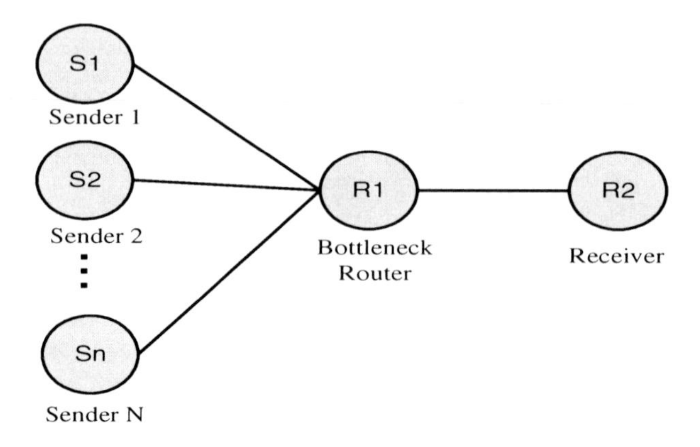


**Fig. 3.54** Network topology

interested in the backbone applications, we consider a single congested router as a representative (similar standpoint can be found in [48, 69]). In this topology, we assume that $N$ TCP connections represent homogeneous and long lived flows. In [69], a fluid-flow model describing the behavior of the average values of key network variables (window size in senders, Round-Trip Time (RTT), and the queue size in the bottleneck router) (Figure 3.54) is given by the following coupled, nonlinear differential equations:

$$\dot{W}(t) = \frac{1}{R(t)} - \frac{W(t)W(t-R(t))}{2R(t-R(t))}p(t-R(t)), \tag{3.63}$$

$$\dot{q}(t) = \frac{W(t)}{R(t)}N(t) - C \tag{3.64}$$

where $R(t) = \frac{q(t)}{C} + T_p$ is the RTT, $W$ is the average TCP window size, $p$ is the dropping probability of a packet, $N$ is the number of connections or TCP sessions, $C$ is the transmission capacity of the router, $q$ is the average queue length of the router buffer, and $T_p$ is the propagation delay. In Equation (3.63), $\frac{1}{R(t)}$ term models *additive increase* of window size, and $\frac{W(t)}{2}$ term models the *multiplicative decrease* of window size in response to packet dropping $p$. Equation (3.64) models the bottleneck queue length. Considering the physical restriction of queue length and window size, we set $\dot{W}(t) = 0,\ if\ W = 0\ and\ \dot{W}(t) < 0$; $\dot{q}(t) = 0,\ if\ q = 0\ and\ \dot{q}(t) < 0$; $\dot{q}(t) = B,\ if\ q = B\ and\ \dot{q}(t) > 0$, where $B$ is the buffer size.

To allow the use of traditional control theory, small-signal linearization was carried out at some equilibrium point $(W_0, q_0, p_0)$. Assuming $N$, $C$, and $R$ to be constants, we can find the equilibrium point defined by $\dot{W}(t) = 0$ and $\dot{q}(t) = 0$, obtaining

$$\left\{ \begin{array}{l} W_0^2 p_0 = 2 \\ W_0 = \frac{RC}{N} \end{array} \right\}. \tag{3.65}$$

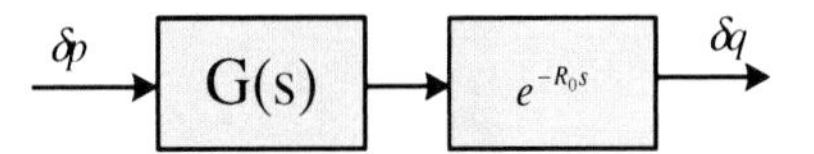


**Fig. 3.55** Linearized system with delay

Linearizing (3.63) and (3.64) around the equilibrium point, one obtains

$$\delta\dot{W}(t) = -\frac{N}{R^2C}(\delta W(t) + \delta W(t-R)) - \frac{R^2C}{2N^2}\delta p(t-R), \tag{3.66}$$

$$\delta\dot{q}(t) = \frac{N}{R}\delta W(t) - \frac{1}{R}\delta q(t) \tag{3.67}$$

where $\delta W(t) \doteq W(t) - W_0$ $\delta q(t) \doteq q(t) - q_0$ $\delta p(t) \doteq p(t) - p_0$ Since, normally, $W \gg 1$ in congestion avoidance stage, we assume $W(t) = W(t-R)$, and simply the Equation (3.66) as

$$\delta\dot{W}(t) = -\frac{2N}{R^2C}\delta W(t) - \frac{R^2C}{2N^2}\delta p(t-R)\,. \tag{3.68}$$

By means of Laplace transform, one can obtain the nominal $s$-domain Linear Time Invariant (LTI) TCP model as

$$G(s) = \frac{\frac{C^2}{2N_0}}{(s + \frac{2N_0}{R_0^2C})(s + \frac{1}{R_0})}\,. \tag{3.69}$$

With the queue dynamics delay, the linearized system is shown in Figure 3.55, where $\delta p$ and $\delta q$ represent perturbed variables about the equilibrium point. This nominal model relates how packet marking probability dynamically affects the queue length.

#### 3.4.4.1 The AQM Control Problem

Before getting into the details of designing an AQM scheme using SSC, we first briefly analyze and implement some popular AQM schemes currently being published, such as RED scheme, adaptive RED scheme, and PI scheme.

**RED Scheme.** The RED scheme uses a low-pass filter to calculate the average queue size, to play down the impact of the burst traffic or transient congestion on the average queue size. Thus, the low-pass filter is designed as an Exponentially Weighted Moving Average (EWMA):

$$avg_q \leftarrow (1 - W_q)avg_q + W_q q \tag{3.70}$$

where $W_q$ is the time constant for the low-pass filter.

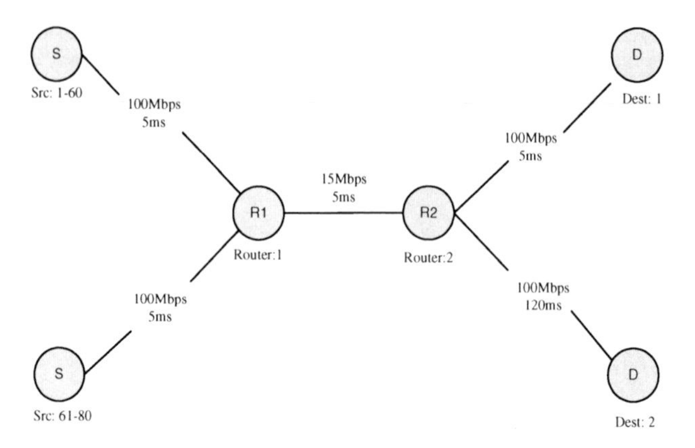


**Fig. 3.56** Simulation setup

$$p(avg_q) = \left\{ \begin{array}{cc} 0, & avg_q < \min_{th} \\ \frac{avg_q - \min_{th}}{\max_{th} - \min_{th}} p_{\max}, & \min_{th} \leq avg_q < \max_{th} \\ 1, & \max_{th} \leq avg_q \end{array} \right\} \tag{3.71}$$

where $\min_{th}$, $\max_{th}$, and $p_{\max}$ are fixed parameters, which are determined in advance by considering the desired bounds on the average queue size.

**Adaptive RED Scheme.** There are several types of Adaptive RED scheme currently available [32], [33]. However, the overall guidelines are the same, that is, adaptively tuning $p_{\max}$ to keep the average queue size between $\min_{th}$ and $\max_{th}$. The main difference between [33] and [32] is that [32] uses multiplicative increase and multiplicative decrease of $p_{\max}$, while [33] uses additive increase and multiplicative decrease of $p_{\max}$.

**PI Scheme.** A classical PI scheme is described in [48]. A generic PI controller, applying to AQM, is

$$p = (k_P + \frac{k_I}{s})(q - r) \tag{3.72}$$

where $k_P$ is the proportional gain, $k_I$ is the integral gain, and $r$ is the reference queue length. Since $p$ represents the probability of dropping a packet, we set $p \in [0, 1]$.

**Simulation Setup.** To show the performance of above schemes, we set up a simple bottleneck network topology with two routers for simulations both in MATLAB and ns-2, as shown in Figure 3.56. In ns-2, the sampling frequency is set as 170 $Hz$. Assume that the transmission capacity of both routers is $C = 15Mbps\,(3750\,packets/s)$, and buffer size is $B = 800\,packets$. At the beginning, $Source: 1-60$ start sending data to $Destination: 1$; at $t = 50\ s$, $Source: 1-60$ switch to send data to

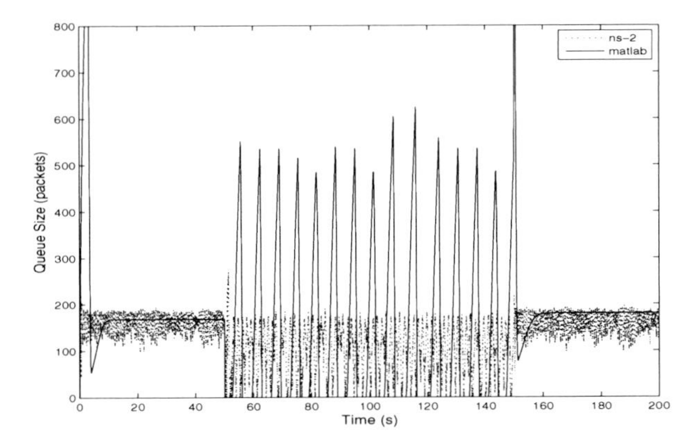


**Fig. 3.57** Instantaneous queue length using RED scheme

*Destination* : 2; at $t = 150\,s$, *Source* : $1-60$ switch back to send data to *Destination* : 1, and *Source* : $61-80$ also start sending data to *Destination* : 1.

1. For RED scheme, the parameters are chosen as $W_q = 0.1$, $\min_{th} = 150\,packets$, $\max_{th} = 250\,packets$, and $p_{\max} = 0.5$.
2. For adaptive RED scheme, we simulate the scheme in [32] with $\alpha = 2.0, \beta = 3.0$. The interval is $0.5\,s$.
3. For PI scheme, the queue length reference $r = 200\,packets$, $K_P = 0.0000182$, and $K_I = 0.00000964$, which is designed at $R_0 = 0.246\,s$ [48], the linear average of the minimum and maximum delays in our simulation setup.

The queue length performance results of RED, adaptive RED, and PI schemes are shown in Figures 3.57, 3.58, and 3.59, respectively. Comparing Figure 3.57 with Figure 3.58, we can find that the oscillation in adaptive RED scheme is relatively smaller than in RED scheme in the small delay period. However, in the large delay period, the oscillation is still quite large in both schemes. Moreover, the queue is empty for a considerable time, which reduces the link utilization. In the PI scheme, the actual queue length cannot track the reference queue length at the first and last 50 seconds partly because the proportional gain is too small compared with the corresponding values at the nominal points. The ns-2 output is to simulate the real network performance. The difference between MATLAB and ns-2 output is because the MATLAB output is based on the simplified fluid-flow model, which assumes that all TCP connections are homogeneous and long lived flows. Moreover, the slow start scheme has not been modeled in the fluid-flow model.

In satellite networks, in particular multi-layer satellite networks, the number of connections and data communication delays can be significantly changed. For example, the round trip time RTT in the communication between LEO (Low Earth Orbit) satellites and MEO (Medium Earth Orbit) satellites is much larger than the delay among LEO satellites. Hence, current RED, adaptive RED, and single PI methods, which are designed based on linearization model at one equilibrium point,

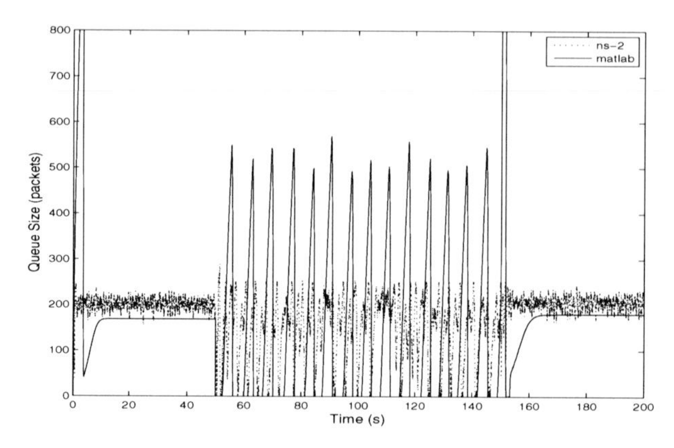


**Fig. 3.58** Instantaneous queue length using adaptive RED scheme

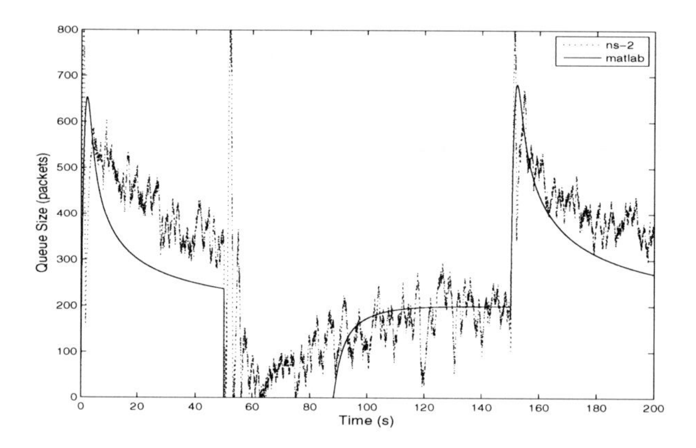


**Fig. 3.59** Instantaneous queue length using PI scheme

are not effective in the multi-layer satellite networks, since both $N$ and $R$ are dynamically varying.

#### 3.4.4.2 Safe Switching Control

The unfalsified SSC system is described below and shown in Figure 3.63, utilizing a set of candidate PID controllers, which is one of the most widely used methods in the industry. Fundamentals of the unfalsified control theory and subsequent extension to the SSC algorithms are described in [20], [54], [94] and references therein. To apply this approach, we first construct a bank of PID controllers. Then, we generate the fictitious reference signal and fictitious error signal for each individual PID controller. Given the fictitious reference signal, plant input signal, and plant output signal sets, the "best" (optimal) controller is selected from the candidate set using a

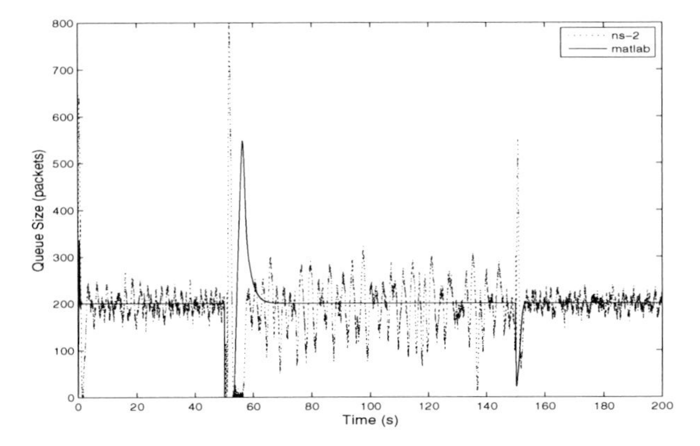


**Fig. 3.60** Instantaneous queue length using SSC scheme

properly designed cost function. To improve the overall performance, the candidate controller parameters can be designed off-line using Integral Squared Error (ISE) optimization algorithms, by considering the linearized models in several general cases [79].

*Problem* 1 *(Candidate Controller Construction)* : The first step of designing unfalsified SSC system is to construct candidate controllers, which are able to detect and control the incipient as well as the current congestion proactively around the corresponding nominal points by regulating the queue length around a preferred queue length reference ($r$). The PID controller is one of the most important control elements in the control industry. A generic PID control equation can be expressed as

$$\begin{aligned} u &= (k_P + \tfrac{k_I}{s} + sk_D)e \\ &= (k_P + \tfrac{k_I}{s})e + sk_D y \end{aligned} \tag{3.73}$$

where $e = y - r$ is the queue length error signal, $r$ is the queue length reference, $u$ is the calculated dropping packet probability, and $y$ is the actual queue length. $k_P$, $k_I$ and $k_D$ are proportional gain, integral gain, and derivative gain respectively. The control output signal $u$ is a combination of current error, the integral of previous errors, and the changing rate of current error. However the generic PID controller in Equation (3.73) is an improper transfer function, and it is hard to exactly implement the derivative part. Hence, the PID controller is written as

$$u = (k_P + \frac{k_I}{s})e + \frac{sk_D}{\varepsilon s + 1}y \tag{3.74}$$

where the parameter $\varepsilon$ is a small number, which is added to approximate the derivative part.

A representative and successful design method using PID controller parameters is to minimize the ISE performance index corresponding to the linearized model [79]:

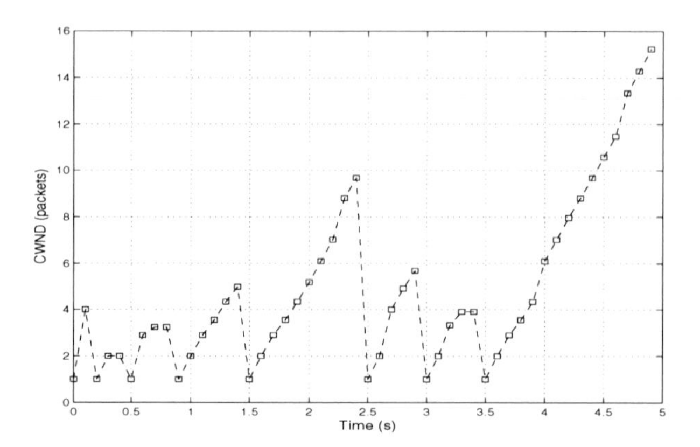


**Fig. 3.61** Congestion window size using SSC scheme

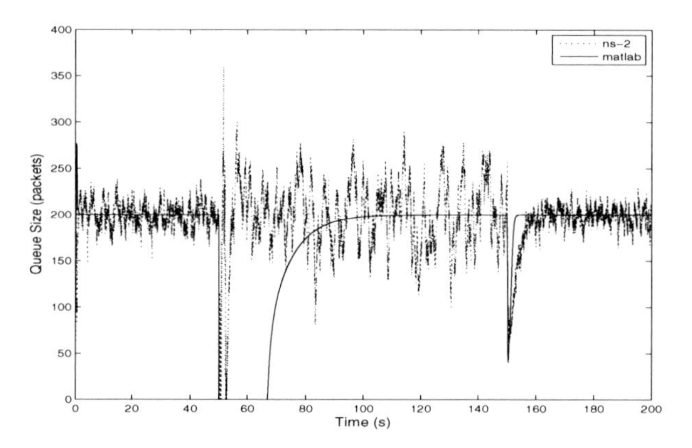


**Fig. 3.62** Instantaneous queue length using SSC scheme with Smith Predictor

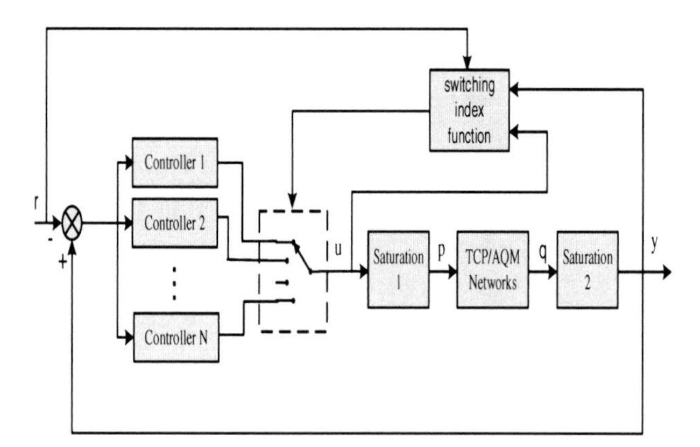


**Fig. 3.63** Unfalsified congestion control structure

$$I = \int_0^{\infty} e^2(t)dt \,. \tag{3.75}$$

Since the control performance optimization is non-convex, a local minimum might occur. To counteract this, the stability margin based on Ziegler-Nichols rules can be used for initial controller parameters [79]. Moreover, the parameters will be examined and adjusted online to achieve satisfied performance before they get selected into the candidate controller set.

For a digital implementation, we need to approximate the integral and the derivative terms to forms suitable for computation by a computer, as,

$$\frac{de(t)}{dt} \approx \frac{e(k) - e(k-1)}{T_s},$$

$$\int_0^t e(t)dt \approx T_s \sum_0^k e(i)$$

where $k = \left\lfloor \frac{t}{T_s} \right\rfloor$, and $T_s$ is the sampling time. Therefore, the digital PID controller becomes

$$u(k) = k_P e(k) + k_I T_s \sum_0^k e(i) + \frac{k_D}{T_s}[e(k) - e(k-1)]$$

where $e(k)$ is the sampled error, and $u(k)$ is the sampled control signal.

*Problem* 2 (*Fictitious Reference Signal*) : Generating fictitious reference signals is an important part in SSC. Given measurements of plant input-output signals $u$ and $y$, there is corresponding fictitious reference signal for each candidate controller, which is a hypothetical signal that would have exactly produced the measured data $(u,y)$, if the $i^{th}$ candidate controller had been in the loop during the entire time period over which the measured data $(u,y)$ was collected. From the Equation (3.74), the fictitious reference signal for the candidate controller $i$ can be calculated as:

$$\tilde{r}_i = y - \frac{s}{sk_{P_i} + k_{I_i}}(u - \frac{sk_{D_i}}{\varepsilon s + 1}y) \,. \tag{3.76}$$

The fictitious error signal for the candidate controller $i$ (defined as the error between its fictitious reference signal and the actual plant output) can be computed as:

$$\tilde{e}_i = y - \tilde{r}_i \,. \tag{3.77}$$

*Problem* 3 (*Cost Function*) : Another important step is to design a suitable cost function to adjust controller parameters based on measured data alone. A standard form of the cost function used in switching control literature contains some form of the accumulated error, such as the cost considered in [11]:

$$J_i(t) = \alpha e_i^2(t) + \beta \int_0^t e^{-\lambda(t-\tau)} e_i^2(\tau)d\tau \tag{3.78}$$

where $\alpha \geq 0$ and $\beta > 0$ can be chosen to yield a desired combination of instantaneous and long-term accuracy measures. The forgetting factor $\lambda > 0$ determines the memory of the index in rapidly switching environments. In [94], it was shown that this cost function is not cost-detectable, and therefore, it may in some cases discard the stabilizing controller and latch onto a destabilizing one. We also compared the performance of cost function (3.78) with (3.83) through simulation in [21].

**Switched System Stability.**
To briefly recall the results on stability in a multi-controller unfalsified setting, consider the system $\Sigma : \mathfrak{L}_{2e} \longrightarrow \mathfrak{L}_{2e}$. We say that stability of the system $\Sigma : \boldsymbol{w} \mapsto z$ is said to be *unfalsified* by the data $(\boldsymbol{w}, z)$ if there exist $\beta, \alpha \geq 0$ such that the following holds:

$$||z||_{\tau} < \beta ||\boldsymbol{w}||_{\tau} + \alpha, \forall \tau > 0 \,. \tag{3.79}$$

Otherwise, we say that stability of the system $\Sigma$ is *falsified* by $(\boldsymbol{w}, \boldsymbol{z})$. In general, $\alpha$ can depend on the initial state. Furthermore, if (3.79) holds with a single pair $\beta,\ \alpha \geq 0$ for all $\boldsymbol{w} \in \mathfrak{L}_{2e}$, then the system is said to be finite-gain stable, in which case the gain of $\Sigma$ is the least such $\beta$.

**Lemma 1.** *( [94]) Consider the switching feedback adaptive control system $\Sigma$, where uniformly bounded reference input $r$, as well as the output $z = [u, y]$ are given. Suppose there are finitely many switches. Let $t_N$ and $K_N$ denote the final switching instant and the final switched controller, respectively. Suppose that the final controller $K_N$ is stably causally left invertible (SCLI) (i.e., the fictitious reference signal $\tilde{r}_{K_N}(z,t)$ is unique and incrementally stable). Then*

$$\left\| \tilde{r}_{K_N} \right\|_t < \|r\|_t + \alpha < \infty, \forall t \geq 0 \tag{3.80}$$

As a switching rule, we consider the cost minimization $\varepsilon$-hysteresis switching algorithm together with the cost functional $J(K, z, t)$. This algorithm returns, at each $t$, a controller $\hat{K}_t$ which is the active controller in the loop:

**$\varepsilon$-hysteresis switching algorithm A1**

[75]

$$\hat{K}_t = \arg\min_{K \in \mathbf{K}} \{J(K, z, t) - \varepsilon \delta_{K\hat{K}_{t^-}}\}$$

where $\delta_{ij}$ is the Kronecker's $\delta$, and $t^-$ is the limit of $\tau$ from below as $t \to \tau$.

The switch occurs only when the current unfalsified cost related to the currently active controller exceeds the minimum (over the finite set of candidate controllers $\mathbf{K}$) of the current unfalsified cost by at least $\varepsilon$. The hysteresis step $\varepsilon$ serves to limit the number of switches on any finite time interval to a finite number, and so prevents the possibility of the limit cycle type of instability. It also ensures a non-zero dwell time between switches.

**Definition 1.** *( [94]) Let $r$ denote the input and $z_d = \Sigma(\hat{K}_t, \mathscr{P})r$ denote the resulting plant data collected while $\hat{K}_t$ is in the loop. Consider the adaptive control system*

*$\Sigma(\hat{K}_t, \mathscr{P})$ with input $r$ and output $z_d$. The pair $(J, \mathbf{K})$ is said to be* cost detectable *if, without any assumption on the plant $P$ and for every $\hat{K}_t \in \mathbf{K}$ with finitely many switching times, the following statements are equivalent:*

1. *$J(K_N, z_d, t)$ is bounded as $t$ increases to infinity.*
2. *Stability of the system $\Sigma(\hat{K}_t, \mathscr{P})$ is unfalsified by the input-output pair $(r, z_d)$.*

**Theorem 1.** *( [94]) Consider the feedback adaptive control system $\Sigma$, together with the hysteresis switching algorithm A1. Suppose the following holds: the adaptive control problem is feasible (there is at least one stabilizing controller in the candidate set), the associated cost functional $J(K, z, t)$ is monotone in time, the pair $(J, \mathbf{K})$ is cost detectable, and the candidate controllers have stable causal left inverses. Then, the switched closed-loop system is stable. In addition, for each $z$, the system converges after finitely many switches to the controller $K_N$ that satisfies the performance inequality*

$$J(K_N, z, \tau) \le J_{true}(K_{RSP}) + \varepsilon \text{ for all } \tau \, . \tag{3.81}$$

**Cost-detectable Cost Function**

As in [94], an example of the cost function and the conditions under which it ensures stability and finiteness of switches according to Theorem 1 may be constructed as follows:

$$J_i(t) = \max_{\tau \in [0,t]} \frac{\varepsilon \left\| u \right\|_\tau^2 + \left\| \tilde{e}_i \right\|_\tau^2}{\left\| \tilde{r}_i \right\|_\tau^2 + \alpha} \tag{3.82}$$

where $u$ is the dropping probability. The weighting parameter $\varepsilon$ is some positive constant. Constant $\alpha$ is used to prevent $J_i(t)$'s denominator to be zero when $\tilde{r}_i = 0$. $\left\| \cdot \right\|_\tau$ stands for the truncated 2-norm $\left\| x \right\|_\tau = \sqrt{\int_0^t (x(\tau))^2 d\tau}$.

However, in satellite TCP/AQM networks, the plant changes when N and R are varying, and the cost function in (3.82) may not be suitable. Since the cost function in (3.82) takes all historical data with the same weight into account, and it is also monotone non-decreasing, the best controller might take a long time to be detected. Hence, the use of this cost function has limitations from the performance standpoint. One approach is to use interval cost function, as (3.83) with detailed description in Algorithm 3.2. Moreover, in the congestion control of TCP/AQM networks, the input $u$ is dropping probability, which is not needed to be penalized. Instead, we restrict the deviations of the dropping probability by penalizing $\Delta u$. As a consequence, the interval cost function is given as follows:

$$J_i(t) = \max_{\tau \in [t_{n_0}, t]} \frac{\varepsilon \left\| \Delta u \right\|_\tau^2 + \left\| \tilde{e}_i \right\|_\tau^2}{\left\| \tilde{r}_i \right\|_\tau^2 + \alpha} \tag{3.83}$$

where $\Delta u$ is the deviation of the dropping probability, and $t_{n_0}$ is the time of the cost function being reactivated at the $nth$ time. The truncated 2-norm $\|x\|_\tau = \sqrt{\int_{t_{n_0}}^{t}(x(\tau))^2 d\tau}$.

**Algorithm 3.2.** 1: Initialization: define a set of candidate controllers, and an initial controller in the loop at the beginning. Set initial cost function output to be 0. Initialize a timer $T = 0s$.

2: Measure $\Delta u$ and $y$. Run the timer.
3: Calculate $\tilde{r}_i$ and $\tilde{e}_i$, and $J_i(t)$.
4: Switch the controller $\arg\min_{1\le i\le N} J_i(t)$ into the loop if $\min_{1\le i\le N} J_i(t) + \varepsilon < J_{current\_controller}(t)$.
5: Measure $y$, and calculate $e' = r^2 - y^2$.
6: If $|e'| > e_{max}$, then initialize the timer to 0, and go back to step 2.
7: If $|e'| \le e_{max}$, then and $T \le t_{max}$, go back to step 2.
8: If $|e'| \le e_{max}$, then and $T > t_{max}$, stop the timer, initialize the cost function output to be 0, and go back to step 5 (shut off the cost monitor).

*Remark* : In Algorithm 3.2, we choose $e_{max} \le r^2(t), \forall t > 0$. With Algorithm 3.2, the cost function in (3.83) is not necessarily monotone non-decrease anymore, which means the cost level will not have to grow up all the time. The historical data, which usually prolong the waiting time of the new "best " controller being switched into the loop, are discarded, according to some designed standard. Therefore, the best controller is detected faster, and performance is improved. Moreover, when the cost monitor is shut off, the computational cost is also saved, which is a consideration of even great importance in satellite networks.

**Lemma 3.4.** *Consider the cost function in (3.83) with $\alpha > 0$. If the candidate controllers in the set $\mathbf{K}$ are in PID form, as shown in (3.74), then $(J, \mathbf{K})$ is cost-detectable*

*Proof.* First, for candidate controllers in PID form, given past values of $u(t)$ and $y(t)$, there always exists a unique fictitious reference signal $\tilde{r}_i$ associated with each controller $K_i$.

Then, assume that the controller switched in the loop at time $t_{n_i}$ is denoted $K_{n_i}$, $t_{0_0} = 0$, and $t_{n_m}$ the time of the final switch of the cost function being turned on at the $nth$ time.

*Step* 1 : Consider the time interval $[t_{0_0}, t_{0_1})$. During this time period, the active controller in the loop is $\hat{K}_t = K_{0_0}$

$$
\begin{aligned}
J(K_{0_0}, z, t_{0_1}^-) &= J(K_{0_0}, z, t_{0_1}) \\
&= \max_{\tau\in[t_{0_0}, t_{0_1})} \frac{\varepsilon \|\Delta u\|_\tau^2 + \left\|\tilde{e}_{K_{0_0}}\right\|_\tau^2}{\left\|\tilde{r}_{K_{0_0}}\right\|_\tau^2 + \alpha} \\
&= \max_{\tau\in[t_{0_0}, t_{0_1})} \frac{\varepsilon \|\Delta u\|_\tau^2 + \|e\|_\tau^2}{\|r\|_\tau^2 + \alpha}
\end{aligned}
$$

since $\tilde{r}_{K_{0_0}} \doteq \tilde{r}(K_{0_0}, z, t) \equiv r(t), t \in [0, t_{0_1})$. Since $r$ is uniformly bounded, $\|r\|_t^2 = \int_{t_{0_0}}^t (r(\tau))^2 d\tau < \infty$. At $t = t_{0_1}$, the cost of the current controller exceeds the current minimum by $\varepsilon$:

$$J(K_{0_0}, z, t_{0_1}) = \max_{\tau \in [t_{0_0}, t_{0_1})} \frac{\varepsilon \|\Delta u\|_\tau^2 + \left\|\tilde{e}_{K_{0_0}}\right\|_\tau^2}{\left\|\tilde{r}_{K_{0_0}}\right\|_\tau^2 + \alpha}$$
$$= \varepsilon + \min_K J(K, z, t_{0_1}) \,. \tag{3.84}$$

Hence, according to the hysteresis switching algorithm, a switch occurs to the controller $K_{0_1} \doteq \arg\min_K J(K, z, t_{0_1})$. Expression in Equation (3.84) is finite since $\varepsilon$ is finite and

$$\min_K J(K, z, t_{0_1}) \leq \sup_{t \in \mathbf{T}, z \in \mathbf{Z}} \min_K J(K, z, t) \doteq J_{true1}(K_{RSP})$$

where $J_{true1}(K_{RSP})$ stands for the true cost in the period in which the cost monitor is active at the $1st$ time, which is finite due to the feasibility assumption. Denoting the sum $\varepsilon + \min_K J(K, z, t_{0_1})$ by $\psi_{0_1}$, we have

$$\max_{\tau \in [t_{0_0}, t_{0_1})} \frac{\varepsilon \|\Delta u\|_\tau^2 + \left\|\tilde{e}_{K_{0_0}}\right\|_\tau^2}{\left\|\tilde{r}_{K_{0_0}}\right\|_\tau^2 + \alpha} = \psi_{0_1}$$
$$\Rightarrow \varepsilon \|\Delta u\|_{t_{0_1}}^2 + \left\|\tilde{e}_{K_{0_0}}\right\|_{t_{0_1}}^2 \leq \psi_{0_1}(\left\|\tilde{r}_{K_{0_0}}\right\|_{t_{0_1}}^2 + \alpha)$$
$$\Rightarrow \varepsilon \|\Delta u\|_{t_{0_1}}^2 + \left\|\tilde{e}_{K_{0_0}}\right\|_{t_{0_1}}^2 \leq \psi_{0_1}(\|r\|_{t_{0_1}}^2 + \alpha) < \infty \,.$$

By induction, while a final switch occurs to the controller $K_{0_m} \doteq \arg\min_K J(K, z, t_{0_m})$, we have

$$\varepsilon \|\Delta u\|_{t_{0_m}}^2 + \left\|\tilde{e}_{K_{0_m}}\right\|_{t_{0_m}}^2 \leq \psi_{0_m}(\left\|\tilde{r}_{K_{0_m}}\right\|_{t_{0_m}}^2 + \alpha) < \infty$$

and $\tilde{e}_{K_{0_m}} = y - \tilde{r}_{K_{0_m}}$. Therefore, by the triangle inequality of norm and Lemma 1, we have

$$\|y\|_{t_{0_m}} \leq \left\|\tilde{e}_{K_{0_m}}\right\|_{t_{0_m}} + \left\|\tilde{r}_{K_{0_m}}\right\|_{t_{0_m}}$$
$$\leq \sqrt{\psi_{0_m}(\left\|\tilde{r}_{K_{0_m}}\right\|_{t_{0_m}}^2 + \alpha)} + \left\|\tilde{r}_{K_{0_m}}\right\|_{t_{0_m}}$$
$$\leq \sqrt{\psi'_{0_m}(\left\|\tilde{r}_{K_{0_m}}\right\|_{t_{0_m}}^2 + \alpha'_{0_m})}$$
$$\leq \sqrt{\psi''_{0_m}(\|r\|_{t_{0_m}}^2 + \alpha''_{0_m})} < \infty$$

for some $\psi''_{0_m}, \alpha''_{0_m} \geq 0$.

*Step* 2 : Assume that, after the final switch in the period in which the cost monitor is active at the 1*st* time, the output converges to the reference signal with $|e| \leq e_{max}$ and $T > t_{max}$ in Algorithm 3.2, and the cost monitor is shut off for time $\delta t_1$ before being turned on again at $t_{1_0}$. Therefore, for $t \in [t_{0_0}, t_{1_0} - \delta t_1)$, we have

$$\max_{\tau \in [t_{0_0}, t_{1_0} - \delta t_1)} \frac{\varepsilon \|\Delta u\|_\tau^2 + \left\|\tilde{e}_{K_{0m}}\right\|_\tau^2}{\left\|\tilde{r}_{K_{0m}}\right\|_\tau^2 + \alpha}$$
$$< \varepsilon + \min_K J(K, z, t_{1_0} - \delta t_1)$$
$$< \varepsilon + J_{true1}(K_{RSP}) < \infty$$

Therefore, for the same reason as in *step* 1, we have,

$$\|y\|_{t_{1_0} - \delta t_1} \leq \sqrt{\psi'''_{0_m}(\|r\|^2_{t_{1_0} - \delta t_1} + \alpha'''_{0_m})} < \infty .$$

For $t \in [t_{1_0} - \delta t_1, t_{1_0})$, by the triangle inequality of absolute value, we have,

$$y^2 \leq e_{max} + r^2$$

$$\Rightarrow \|y\|^2_{(t_{1_0} - \delta t_1, t_{1_0})} \leq e_{max} \delta t_1 + \|r\|^2_{(t_{1_0} - \delta t_1, t_{1_0})} .$$

Since $e_{max} \leq r^2(t), \forall t > 0$, we have

$$\|y\|^2_{(t_{1_0} - \delta t_1, t_{1_0})} \leq 2\|r\|^2_{(t_{1_0} - \delta t_1, t_{1_0})}$$

*Step* 3 : Consider the time interval $[t_{1_0}, t_{1_1})$. During this time period, the active controller in the loop is $\hat{K}_t = K_{1_0}$. Denoting the sum $\varepsilon + \min_K J(K, z, t_{1_1})$ by $\psi_{1_1}$. Similar with *step* 1, we can conclude that

$$\varepsilon \|\Delta u\|^2_{(t_{1_0}, t_{1_1})} + \left\|\tilde{e}_{K_{1_0}}\right\|^2_{(t_{1_0}, t_{1_1})} \leq \psi_{1_1}(\left\|\tilde{r}_{K_{1_0}}\right\|^2_{(t_{1_0}, t_{1_1})} + \alpha)$$

$$\Rightarrow \|y\|_{(t_{1_0}, t_{1_1})} \leq \sqrt{\psi'_{1_1}(\|r\|^2_{(t_{1_0}, t_{1_1})} + \alpha'_{1_1})} < \infty$$

and so as $\|y\|_{(t_{1_0}, t_{1_m})}$ and $\|y\|_{(t_{1_0}, t_{2_0} - \delta t_2)}$.

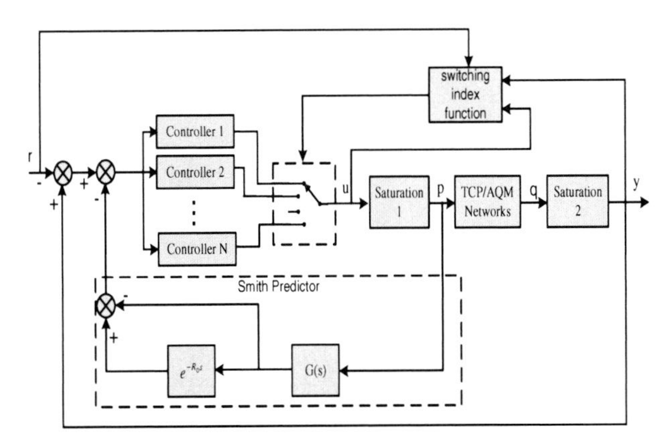


**Fig. 3.64** Unfalsified congestion control structure with Smith Predictor

Finally, by summing up all $\|y\|_\tau^2$ in above steps, we can conclude that

$$\begin{aligned}
\|y\|_t^2 &= \|y\|_{t_{1_0}-\delta t_1}^2 + \|y\|_{(t_{1_0}-\delta t_1, t_{1_0})}^2 + \|y\|_{(t_{1_0}, t_{2_0}-\delta t_2)}^2 + \cdots \\
&\leq \psi'(\|r\|_{t_{1_0}-\delta t_1}^2 + \|r\|_{(t_{1_0}-\delta t_1, t_{1_0})}^2 + \\
&\quad \|r\|_{(t_{1_0}, t_{2_0}-\delta t_2)}^2 + \cdots) + \alpha' \\
&= \psi' \|r\|_t^2 + \alpha', \ \forall t > 0 \\
\Rightarrow \quad & \|y\|_t \leq \psi'' \|r\|_t + \alpha'', \ \forall t > 0
\end{aligned}$$

for some $\psi'', \alpha'' \geq 0$.

From the above, we conclude the stability of the closed-loop switched system, and cost-detectability based on Definition 1.

*Problem* 4 (*Restriction*) : Since the controller output is the probability of dropping the packets, and the plant output is the queue length, there are restrictions for these two variables, shown as follows:

$$p = \min(max(u,0),1) \tag{3.85}$$

$$y = \min(max(q,0),B) \tag{3.86}$$

where $B$ is the total buffer size.

*Problem* 5 (*Smith Predictor*) : In the simulation result(Figure 3.60), it can be found that, though the state finally becomes stable, the overshoots are high, and the settling times are long, when the propagation delay is large, which happens when satellites on different layers are communicating with each other. The Smith Predictor, as shown in Figure 3.64, is well known to be effective to reduce non-preferred

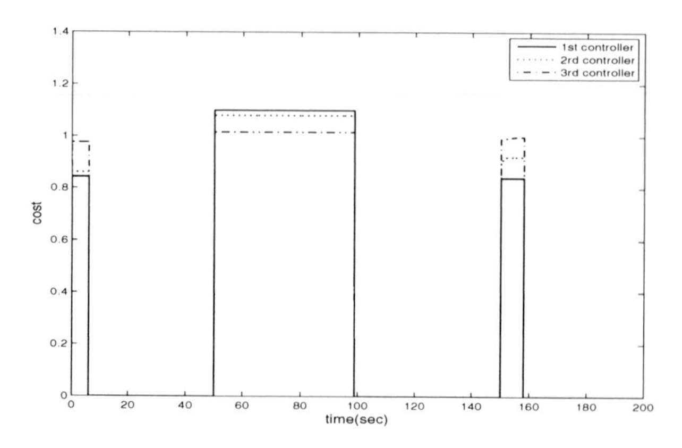


Fig. 3.65 Outputs of cost-detectable cost function (3.83)

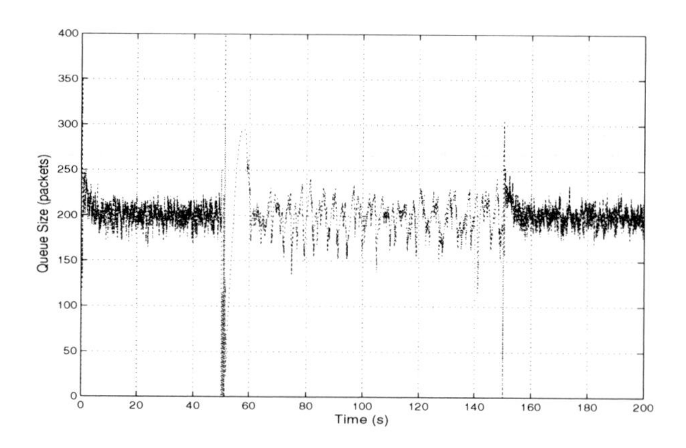


Fig. 3.66 Instantaneous queue length using SSC scheme with TCP NewReno

overshoots and settling times. From previous statement, we know, if the queue length is short, then propagation delay contributes to most RTT. Therefore, the nominal model with RTT equal to round trip propagation delay is very close to actual model, and so we use the round trip propagation time as nominal delay in the smith predictor in this paper.

#### 3.4.4.3 Simulation Results

In this section, we simulate the SSC scheme using the same simulation setup as used in section 3 using MATLAB and ns-2 (the SSC with interval cost function switching algorithm is implemented in matlab). The queue length reference $r = 200\,packets$. Based on the linearized models about the nominal points $N = 60, R = 0.03$ and $N = 60, R = 0.26$, the candidate controller set is designed off-line, using

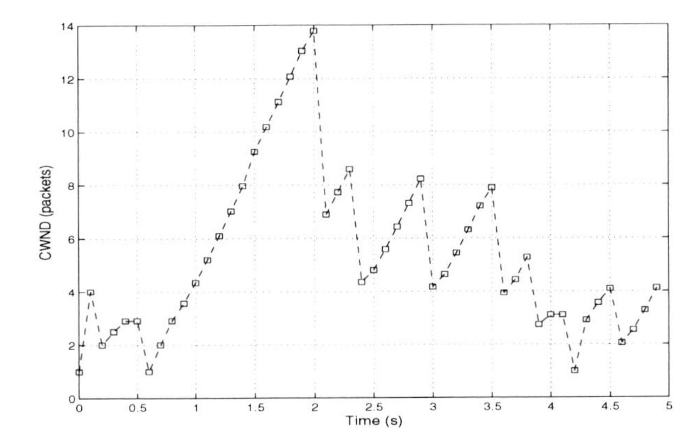


**Fig. 3.67** Congestion window size using SSC scheme with TCP NewReno

ISE optimization algorithm [79], as follows. The parameters for the first controller are $k_P = 0.000951, k_I = 0.0032$, and $k_D = 0.00004993$, those for the third controller $k_P = 0.000047, k_I = 0.000021081, k_D = 0.0000098984$. We also introduce a second controller $k_P = 0.000499, k_I = 0.0016, k_D = 0.000029914$ as a linear interpolation of the first and third controllers (the average of their parameters). In Algorithm 3.2, we assign $e_{max} = 2000\,packets^2$ and $t_{max} = 5\,s$.

**Preliminary Simulations with Original TCP.** The instant queue length results are shown in Figures 3.60 and 3.62. The cost output is shown in Figure 3.73. By comparing Figures 3.57, 3.58, and 3.59 with Figures 3.60 and 3.62, we observe that using RED, adaptive RED, or single PI controller does not result in satisfactory tracking the queue length reference. Instead, the proposed SSC unfalsified controller can track the reference signal much better. Even when the number of connections rises by 20, and there is no controller in this nominal point in the candidate set, the proposed cost function can still choose the best controller (best according to the switching algorithm A1), and track the reference signal, which demonstrates the robustness of the SSC unfalsified control system. In Figure 3.73, the time intervals when all the cost function outputs drop to 0 signify the case that $|e^{'}| \leq e_{max}$ and $T > t_{max}$ (in MATLAB), so the cost monitor has been shut off to save the computational cost, and the historical data are discarded to re-initialize the cost level. The change of cost function outputs from zero to nonzero signifies the case that $|e^{'}| > e_{max}$, and the current candidate controller might be falsified. Hence, the cost monitoring is reactivated to detect the optimal controller. The congestion window size of one TCP flow is zoomed in and shown in Figure 3.61. In the original TCP scheme, each time when a packet loss is taken as a sign of congestion. Then the threshold value is set as half of the current window size, the congestion window size is set to one, and slow start begins until the congestion window size reaches the threshold value. The problem is that the original TCP has to wait for timeout to perform a retransmission of the lost segment, which might lead to idle periods. Moreover, each time when a packet loss is indicated, it employs the slow start algorithm to recover from congestion. Sometimes, it is not necessary, especially in satellite networks with long delay. Even though the congestion window size exponentially grows in the slow start procedure, the time it takes to reach the threshold value is still notable because of the long delay and large congestion window size.

**Simulations with TCP NewReno.** To further improve the performance, TCP NewReno, rather than original TCP, is used with the same controller designed above. The main modification of NewReno is the introduction of Fast-Retransmit and Fast-Recovery, which introduces two indications of a packet loss: Retransmission Time-out (RTO) and arrival of three duplicate acknowledgements. If a packet loss is indicated by a timeout, then TCP believes that the network is congested and hence enters into the slow start procedure to recover from it. Arrival of three duplicate acknowledgements means that there are data still flowing between the two ends, and congestion does not happen. Hence TCP employs the Fast-Retransmit and Fast-Recovery to avoid reducing the flow abruptly. Actually, TCP NewReno is even closer to the fluid model in Equation (3.63), which ignores the slow start procedure. Hence, we can see better performance about instant queue length, shown in Figure 3.66. The congestion window size of one TCP flow is zoomed in and shown in Figure 3.67.

#### 3.4.4.4 Simulation with Non-cost Detectable Cost Function

Finally, it is illustrative to compare the performance of the previously described switching control algorithm with the algorithm that uses a non-cost detectable cost function, such as the one commonly used in the literature, shown in (3.87) below.

The simulation parameters are chosen similar as those in the previous section. The plant (satellite network) is described as having the transmission capacity of the router of $C = 15Mbps\,(3750packets/s)$ and the queue length reference $r = 200$ packets. At the beginning, the number of connections is $N = 60$ with the propagation delay $T_p = 0.03\,s$; after 50 seconds, $T_p$ rises up to 0.5 seconds with $N$ unchanged; after 100 seconds, $T_p$ falls back to 0.03 with $N$ rising to 80. Based on the linearized models about the nominal points $N = 60, R = 0.08$ and $N = 60$, $R = 0.5$, the candidate controller set is designed off-line, using integral square error (ISE) optimization algorithm, as follows. The parameters for the first controller are $K_P = 0.000951, K_I = 0.0032$, and $K_D = 0.00004993$, those for the third controller $K_P = 0.000014, K_I = 1.6784 \cdot 10^{-6}$, and $K_D = 0.00000639$. Also introduced is a second controller $K_P = 4.8250 \cdot 10^{-4}, K_I = 0.0016$, and $K_D = 2.8160 \cdot 10^{-5}$ as a linear interpolation of the first and third controllers (the average of their parameters).

$$J_i(t) = -\rho + \int_0^t \Gamma_{spec}(\widetilde{r_i}(t), y(t), u(t))dt \tag{3.87}$$

where $\rho$ is an arbitrary positive number, which is used to judge whether a certain controller is falsified or not, and $\Gamma_{spec}$ is chosen as

$$\Gamma_{spec}(\widetilde{r}_i(t), y(t), u(t)) =$$
$$(w_1 * (\widetilde{r}_i(t) - y(t))^2 + (w_2 * u(t))^2 - \delta^2 - \widetilde{r}_i(t)^2$$

where $w_1$ and $w_2$ are weighting filters chosen by the designer, and $\sigma$ is a constant representing the r.m.s. effects of noise on the cost.

The actual queue length is shown in Figure 3.68, the throughput in Figure 3.69. From Figure 3.70, we see that all controllers are falsified.

Finally, the simulation of the safe switching adaptive algorithm with a cost-detectable cost function is performed. The actual queue length is shown in Figure 3.71, and the throughput is shown in Figure 3.72, whereas the cost output is shown in Figure 3.73. The round trip time time variation is shown in Figure 3.74. By comparing Figures 3.68, 3.69 with Figures 3.71, 3.72, we observe that using one fixed controller or using switching unfalsified control method with the standard (non-cost detectable) cost function does not result in satisfactory tracking the queue length reference, and the bandwidth utility is low. Instead, the proposed safe switching adaptive unfalsified controller can track the reference signal, stabilize the system, and achieve high bandwidth utility in different situations. Even when the number of connections rises by 20, and there is no controller in this nominal point in the candidate set, the proposed cost function can still choose the best controller (best according to the switching algorithm), and track the reference signal, demonstrating its robustness. Figure 3.71 in particular demonstrates the improvement achieved using the cost-detectable switching controller. It shows that the queue length is kept low which is very good from the router perspective. Figure 3.74 shows that the unfalsified controller can keep the RTT at a relatively low level. In Figure 3.73, the time intervals when all the cost functions drop to 0 signify the case when the queue length error with the currently active controller has been lower than five packets for 5 seconds, and so the cost monitor has been shut off to save the computational cost. When the cost functions become non-zero again, it signifies the case when the current error is larger than five packets, and the current candidate controller might be falsified. Hence, the cost monitoring is reactivated to choose the optimal controller.

### *3.4.5 Unfalsified Direct Adaptive Control of a Two-Link Robot Arm*

The safe switching control theory based on controller unfalsification ideas has been successfully applied to the problem of robust adaptive control of a robot manipulator in [99]. It is shown how a priori mathematical knowledge can be merged with data to design a robust adaptive controller.

The dynamics model of an ideal rigid-link manipulator is given in the following form:

$$H(\theta,q)\ddot{q} + C(\theta,q,\dot{q})\dot{q} + g(\theta,q) = u_a \tag{3.88}$$

in which $q$ is a real $n$-dimensional vector representing the rotational angles of the $n$ links of the manipulator arm, $H(\theta,q)$ is the inertia matrix, $C(\theta,q,\dot{q})\dot{q}$ accounts for the joint friction, coupling Coriolis and centripetal forces, respectively, $g(\theta,q)$ is

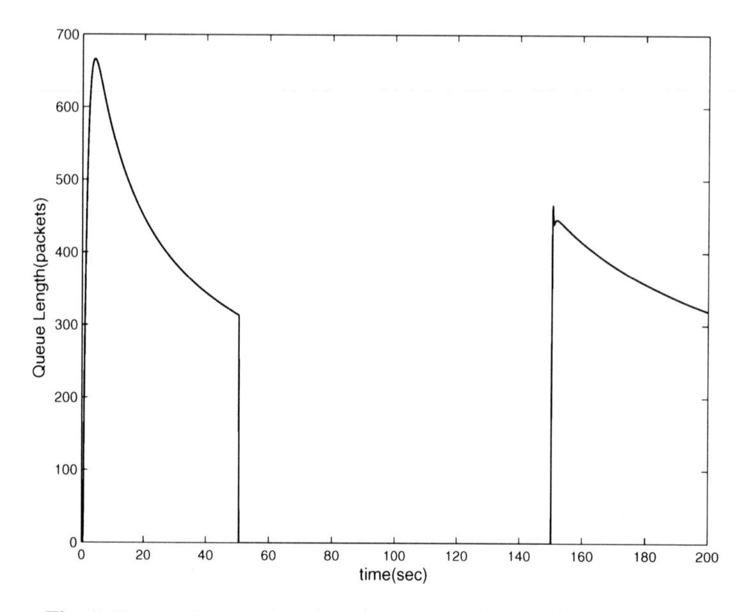


**Fig. 3.68** Actual queue length with a non cost detectable cost function

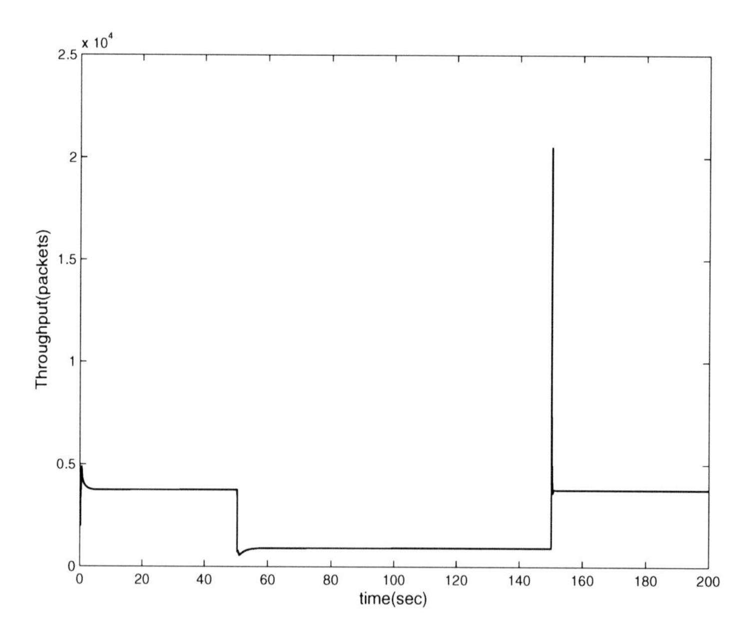


**Fig. 3.69** Throughput with a non cost detectable cost function

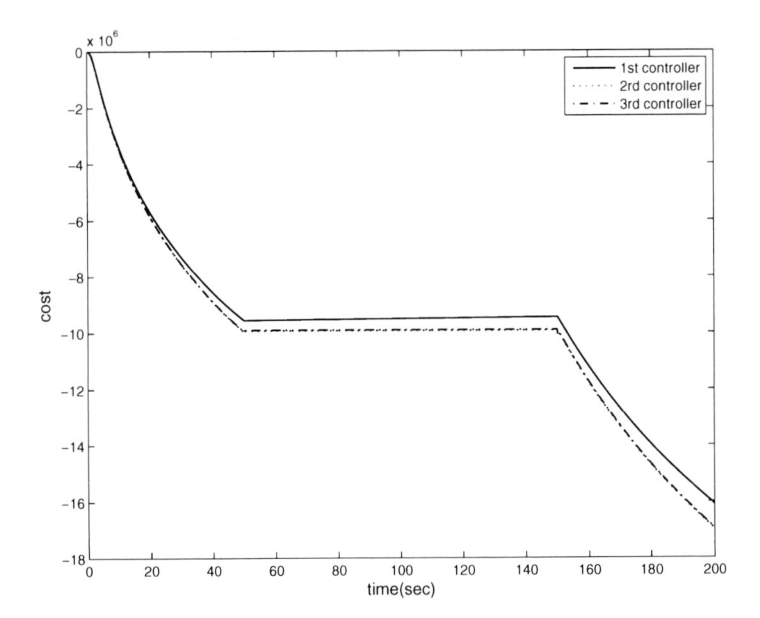


**Fig. 3.70** Outputs of a non cost detectable cost function

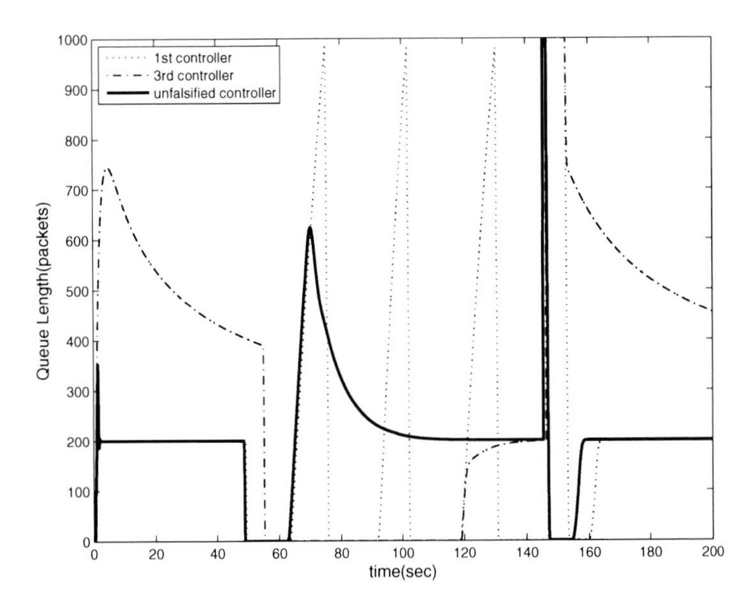


**Fig. 3.71** Actual queue length using safe adaptive control algorithm

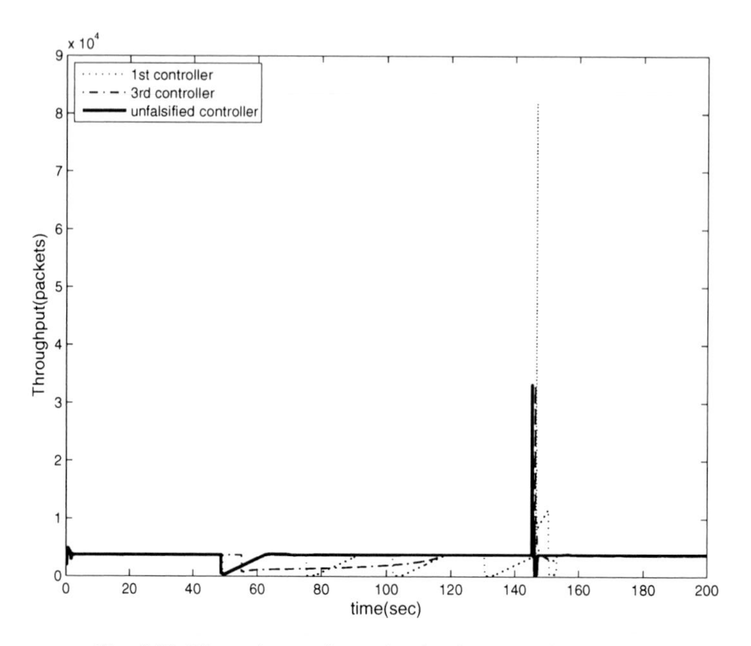


**Fig. 3.72** Throughput using safe adaptive control algorithm

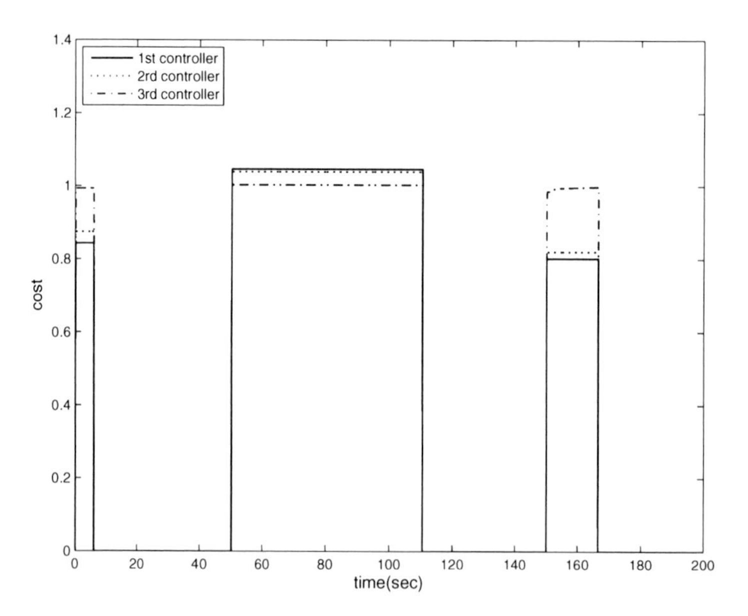


**Fig. 3.73** Cost-detectable cost function outputs

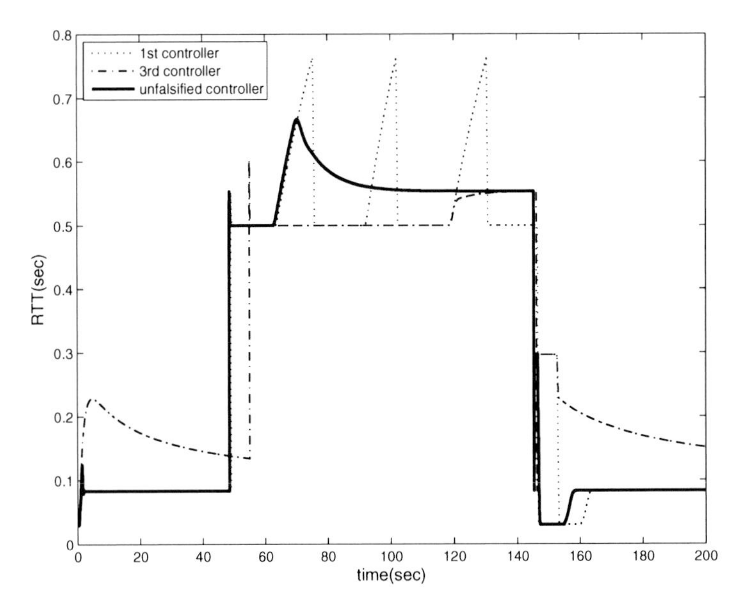


**Fig. 3.74** Round-trip time

the torque caused by gravity, and $u_a$ is a real $n$-dimensional vector whose elements are joint torques consisting of actuator outputs and external disturbances. Physical nature of the problem ensures that the $H(\theta,q)$ is always a positive-definite matrix.

For example, the rigid body dynamics of the planar, two-link manipulator of Slotine and Li [92], shown in Figure 3.75, can be written in the form of (3.88) with

$$H(\theta,q) = H(\theta,q)^T = \left[\begin{array}{c|c} H_{11} & H_{12} \\ \hline H_{21} & H_{22} \end{array}\right]$$

$$C(\theta,q,\dot{q}) = \left[\begin{array}{c|c} -h\dot{q}_2 & -h(\dot{q}_1+\dot{q}_2) \\ \hline h\dot{q}_1 & 0 \end{array}\right]$$

where

$$\begin{aligned} H_{11}(\theta) &= \theta_1 + 2\theta_3 \cos q_2 + 2\theta_4 \sin q_2 \\ H_{12}(\theta) = H_{21}(\theta) &= \theta_2 + \theta_3 \cos q_2 + \theta_4 \sin q_2 \\ H_{22}(\theta) &= \theta_2 \\ h(\theta) &= \theta_3 \sin q_2 - \theta_4 \cos q_2 \\ \theta &= [\theta_1,\ \theta_2,\ \theta_3,\ \theta_4]^T \end{aligned} \tag{3.89}$$

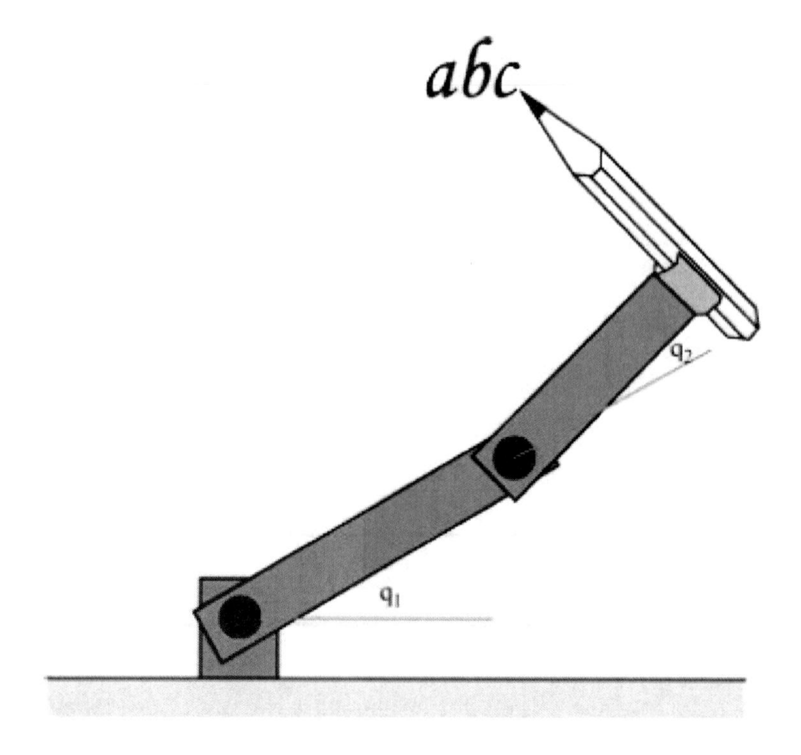


**Fig. 3.75** A two-link robot manipulator with joint angles $\dot{q}_1(t)$ and $\dot{q}_2(t)$

and

$$\begin{aligned} q = [q_1,\ q_2]^T\ ,\ \ u_a = [u_{a1},\ u_{a2}]^T \\ \theta_1 = I_1 + m_1 l_{c1}^2 + I_e + m_e l_{ce}^2 + m_e l_1^2 \\ \theta_2 = I_e + m_e l_{ce}^2 \\ \theta_3 = m_e l_1 l_{ce} \cos\delta_e \\ \theta_4 = m_e l_1 l_{ce} \sin\delta_e\ . \end{aligned} \tag{3.90}$$

The parameters with subscript 1 are related to link 1, while parameters with subscript $e$ are related to the combination of link 2 and end-effector. It is assumed that the manipulator moves in the horizontal plane only, so that the gravity term is zero.

Using the above equations, the manipulator dynamics can take an alternative form:

$$Y(q,\dot{q},\ddot{q})\theta + g(\theta,q) = u_a \tag{3.91}$$

in which $\theta = [\theta_1,\ \theta_2,\ \theta_3,\ \theta_4]^T$ and $Y(\cdot)$ is a $2\times 4$ matrix with elements

$$\begin{aligned} Y_{11} &= \ddot{q}_1,\ Y_{12} = \ddot{q}_2,\ Y_{21} = 0,\ Y_{22} = \ddot{q}_1 + \ddot{q}_2 \\ Y_{13} &= (2\ddot{q}_1 + \ddot{q}_2)\cos q_2 - (2\dot{q}_2\dot{q}_1 + \dot{q}_2^2)\sin q_2 \\ Y_{14} &= (2\ddot{q}_1 + \ddot{q}_2)\sin q_2 + (2\dot{q}_2\dot{q}_1 + \dot{q}_2^2)\cos q_2 \\ Y_{23} &= \ddot{q}_1\cos q_2 + \dot{q}_1^2)\sin q_2 \\ Y_{24} &= \ddot{q}_1\sin q_2 + \dot{q}_1^2)\cos q_2 \end{aligned} \tag{3.92}$$

which makes the dynamical equations linear in the controller parameter vector $\theta$ for the above two-link example.

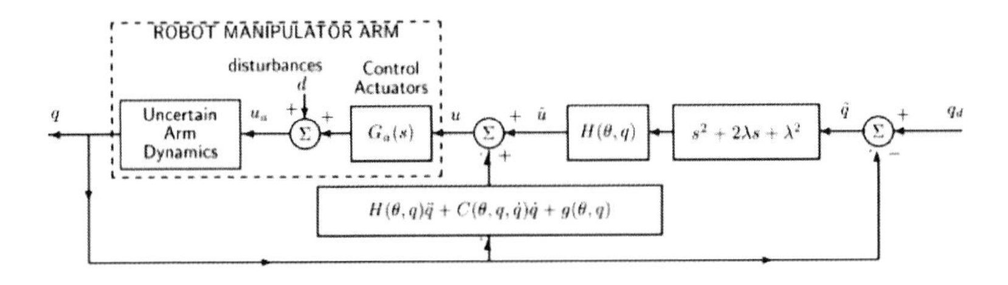


**Fig. 3.76** Computed torque manipulator control configuration

For the manipulator trajectory control, the 'computed torque' control method is commonly used to deal with the non-linearity of the dynamic equation. Figure 3.76 shows an example of a computed torque control law:

$$u = H(\theta,q)[\ddot{\tilde{q}} + 2\lambda\dot{\tilde{q}} + \lambda^2\tilde{q}] + H(\theta,q)\ddot{q} + C(\theta,q,\dot{q})\dot{q} + g(\theta,q) \tag{3.93}$$

$$= \tilde{u} + Y(q,\dot{q},\ddot{q})\theta + g(\theta,q) \tag{3.94}$$

where

$$\tilde{u} = H(\theta,q)[\ddot{\tilde{q}} + 2\lambda\dot{\tilde{q}} + \lambda^2\tilde{q}] \tag{3.95}$$

$$\tilde{q} = q_d - q \tag{3.96}$$

The above computed torque control law does not actually require acceleration measurements. This is seen when the control law is simplified as

$$u = H(\theta,q)[\ddot{q}_d - 2\lambda\dot{\tilde{q}} - \lambda^2\tilde{q}] + C(\theta,q,\dot{q})\dot{q} + g(\theta,q) \tag{3.97}$$

where $q_d$ denotes the desired trajectory, $\tilde{q}$ is the tracking error, and $\lambda > 0$ is a design parameter which determines the speed at which the tracking error converges to zero. The actual joint torque is related to the control signal $u$ by:

$$u_a = G_a(s)u + d \tag{3.98}$$

where $G_a(s)$ represents uncertain actuator dynamics, and $d$ is an uncertain disturbance. In the case of no disturbances, no actuator dynamics ($G_a(s) = 1$ and no modeling uncertainty, we have then $u = u_a$, and the control law 3.93 applied to the ideal manipulator yields:

$$H(\theta, q(t))[\ddot{\tilde{q}} + 2\lambda\dot{\tilde{q}} + \lambda^2\tilde{q}] = 0 \tag{3.99}$$

which, combined with the fact that $H(\theta, q)$ is strictly positive definite for all $q$, implies that the tracking error $\tilde{q}$ exponentially converges to zero with the rate $\lambda$. When an external disturbance is present, the closed loop dynamics become:

$$H(\theta, q(t))[\ddot{\tilde{q}} + 2\lambda\dot{\tilde{q}} + \lambda^2\tilde{q}] = -d(t) \tag{3.100}$$

which indicates that the tracking error $\tilde{q}$ will converge to a region whose size is proportional to the magnitude of the disturbance $d$. These well-known results hold for the idealized case where the parameters are exactly known, the actuators have no dynamics, there is no friction, and the links are completely rigid. As good performance may still be possible when these idealized assumptions do not hold (at least for some values of the assumed parameter vector $\theta$), alternative stability proofs are needed. This is what highlights the utility of the unfalsified, safe switching control theory, which provides a rapid and precise means for determining which, if any, of values of $\theta$ remain suitable for control of the actual non-idealized physical system, based on a real-time analysis of evolving real-time plant data.

In the following, the *a priori* mathematical knowledge and the a posteriori data are combined in the context of unfalsified control theory to produce a robust adaptive controller. In the unlikely event that the manipulator conforms exactly to the theoretical ideal so that its dynamics are exactly described by (3.88) with known parameters, and joint torque is exactly as commanded (*i.e.* joint actuator transfer function $G_a(s) = 1$), then the application of the control law (3.93) will yield satisfactory performance. However, a real physical manipulator will have many other factors that cannot be characterized by (3.88) such as link flexibility and the effects of actuator dynamics, saturation, friction, mechanical backlash and so forth. A mathematical model is never able to describe every detail of a physical system, and so there is always a gap between the model and reality. Such a gap may sometimes be fortuitously bridged when the aforementioned factors are ‘negligible’, but unfalsified control theory provides a more robust methodology, which ensures that this gap will be overcome whenever possible. In the following, the real-time data are directly used to quickly and accurately assess the appropriateness of various control laws of the form (3.93) on a given physical manipulator. The following scenario is assumed:

1. Prior Knowledge: The mathematical knowledge about manipulators in general and prior observation of the particular manipulator’s characteristics indicate that the use of a control law of the general form (3.93) could result in the performance described by (3.100), and that (3.88) and (3.91) should hold.

2. Uncertainty: Parameters such as inertia, location of mass center and so forth cannot be correctly known in advance, because of possible changes of operating conditions or load mass, or because of other causes. Also, there may be other sources of non-parametric uncertainty, such as time-delays, link bending modes, noise/disturbances, actuator dynamics and so forth.
3. Data: The actuator input commands $(u_1, u_2)$ and the manipulator's output angles $(q_1, q_2)$, velocities $(\dot{q}_1, \dot{q}_2)$, and accelerations $(\ddot{q}_1, \ddot{q}_2)$ are directly measurable.

The reference signal $r$, measurement signal $y$ and control input signal $u$ are taken to be:

$$r = q_d, \tag{3.101}$$

$$y = [q_1, q_2, \dot{q}_1, \dot{q}_2, \ddot{q}_1, \ddot{q}_2]^T, \tag{3.102}$$

$$u = [u_1, u_2]^T \tag{3.103}$$

and, at each time $\tau$, the measurement data are

$$u_{data} = P_\tau u \tag{3.104}$$

$$y_{data} = P_\tau y . \tag{3.105}$$

The set of admissible control laws and performance specification are selected as follows:

$$\mathbf{K} = \{\mathbf{K}(\theta) | \theta \in \mathbb{R}^m\} \tag{3.106}$$

$$with \ \mathbf{K}(\theta) = \{(r, y, u) | u = K_\theta(r, y)\}, \tag{3.107}$$

$$T_{spec}(\theta) = \{(r, y, u) | J_\theta(r(t), y(t), u(t)) \leq 0 \ \forall t \leq \tau\} \tag{3.108}$$

where

$$K_\theta(r, y) \triangleq H(\theta, q)[\ddot{q}_d - 2\lambda\dot{\tilde{q}} - \lambda^2\tilde{q}] + C(\theta, q, \dot{q})\dot{q} + g(\theta, q) \tag{3.109}$$

$$J_\theta(r(t), y(t), u(t)) \triangleq |\tilde{u}| - \bar{d} \tag{3.110}$$

where, in (3.110), $\bar{d}(t) \geq 0$ is a given function of time, and in (3.109), $u = u(\theta, q, \dot{q}, q_d, \dot{q}_d, \ddot{q}_d)$, $\tilde{u} = \tilde{u}(\theta, q, \tilde{q}, \dot{\tilde{q}}, \ddot{\tilde{q}})$, and $\tilde{q} = q_d - q$ are previously defined. The inequality $J_\theta(r(t), y(t), u(t)) \leq 0$ means that each entry of the vector $J_\theta(r(t), y(t), u(t))$ is $\leq 0$. Based on the data condition, the measured data $(u_{data}, y_{data})$ consists of past values of commanded joint control torques $u$ and sensor output signals $q, \dot{q}, \ddot{q}$, respectively. In this case, the measurement information set $P_{data}$ is given in terms of the data $(u_{data}, y_{data})$ by

$$P_{data} \triangleq \{(r, y, u) | P_\tau u = u_{data}, P_\tau y = y_{data}\} . \tag{3.111}$$

Denoting the set of unfalsified values of $\theta$ at time $\tau$ as $\Theta(\tau)$, each element $\theta \in \Theta(\tau)$ corresponds to a control law $\mathbf{K}(\theta)$ given by (3.109). The unfalsified controller parameter set $\Theta(t)$ can be obtained as follows. A control law $\mathbf{K}_\theta$ having parameter vector $\theta$ results in the control signal $u$ given by the computed torque control law

3.109. Hence, the set $P_{data} \cap \mathbf{K}(\theta)$ consists of those points $(r, y, u)$ satisfying for all $t \leq \tau$

$$r(t) = \hat{q}_d(\theta)(t) \tag{3.112}$$
$$y(t) = y_{data}(t) \tag{3.113}$$
$$u(t) = u_{data}(t) \tag{3.114}$$

where, for each $\theta$, $\hat{q}_d(\theta)$ is a solution to $u_{data} = K_{data}(\hat{q}_d(\theta), y_{data}$, *viz*,

$$\ddot{\hat{q}}_d(\theta) + 2\lambda\dot{\hat{q}}_d(\theta) + (\lambda)^2\hat{q}_d(\theta) = \tag{3.115}$$
$$= [H(\theta,q)]^{-1}(u + H(\theta,q)(2\lambda\dot{q} + (\lambda)^2 q) - C(\theta,q,\dot{q})\dot{q} - g(\theta,q)) \tag{3.116}$$

Given only the measured plant input-output data and a parameter vector $\theta$, the right-hand side of the above equation can be determined. One may then use this fictitious reference signal $r_0(\theta) \triangleq \hat{q}_d(\theta)$ to test each candidate control law in the feedback loop. Then, the unfalsified controller parameter set $\Theta(t)$ at the current time $\tau$ can be expressed as a set intersection:

$$\Theta(t) = \bigcap_{0 \leq t \leq \tau} \Omega(t) \tag{3.117}$$

where

$$\Omega(t) \triangleq \{\theta | |\tilde{u}(q(t), \dot{q}(t), \ddot{q}(t), \theta)| \leq \bar{d}(t), \theta \in \Theta_0\} \tag{3.118}$$
$$\tilde{u}(q(t), \dot{q}(t), \ddot{q}(t), \theta) \triangleq Y(q(t), \dot{q}(t), \ddot{q}(t))\theta - u(t) \,. \tag{3.119}$$

In 3.119, the set $\Theta_0 \subset \mathbb{R}^m$ is an a priori estimate of the range of controller parameter vectors that are considered to be candidates for achieving the performance specification $T_{spec}(\theta)$.

A controller parameter vector $\hat{\theta}(\tau)$ is to be computed such that $\hat{\theta}(t) \in \Theta(\tau)$, $\forall \tau$. The corresponding controller $K_{\hat{\theta}}(r, y, u)$ may then be inserted in the control loop. The strategy for choosing $\hat{\theta}(\tau)$ is as follows: the value of $\hat{\theta}(\tau)$ is held constant until such time when it is falsified by the latest data. That is, $\hat{\theta}(\tau)$ remains constant as time $\tau$ increases until such time occurs when $\hat{\theta}(\tau^-) \notin \Omega(\tau)$. (Here $\tau^-$ denotes the time just an instant prior to time $\tau$.) At the instant $\tau$ when this occurs $\hat{\theta}(\tau^-) \notin \Theta(\tau)$, and $\hat{\theta}(\tau)$ must switch to a new value $\hat{\theta}(t) \in \Theta(\tau)$.

An important theoretical point to note is that each of the sets $\Omega(\tau)$ defined in (3.119) is a convex polytope, bounded by the intersection of two pairs of hyperplanes in $\mathbb{R}^4$. Also, the intersection of finitely many convex polytopes is a convex polytope, too. The computation of an element of a convex polytope is a linear programming problem for which there are many good computational algorithms. Thus, the computation of an unfalsified $\hat{\theta}(t) \in \Theta(\tau)$ at a switching time involves using a linear programming algorithm.

A parameter update law for $\hat{\theta}$ is proposed such that it produces new controller parameter vector that is 'optimal' in the sense that it is as far as possible from the boundary of the current unfalsified set $\Theta(\tau)$, *i.e.*

$$\hat{\theta}(\tau) = \arg \max_{\theta \in \Theta(\tau)} \text{dist}(\theta, \delta\Theta(\tau)) \tag{3.120}$$

where $\delta\Theta(\tau)$ denotes the boundary of the set $\Theta(\tau)$. Specifically, one computes $\hat{\theta}$ as the solution to the following linear programming problem:

$$\hat{\theta}(\tau) = \arg \max_{\theta \in \Theta(\tau)} \delta \tag{3.121}$$

subject to $\forall 0 \leq t \leq \tau$

$$\begin{aligned} \delta &\geq 0 \\ -Y(q(t),\dot{q}(t),\ddot{q}(t))\theta + \bar{d}(t) + \delta R(t) &\geq 0 \\ Y(q(t),\dot{q}(t),\ddot{q}(t))\theta - \bar{d}(t) - \delta R(t) &\geq 0 \end{aligned} \tag{3.122}$$

where $R(t) \in \mathbb{R}^2$ is given by

$$R(t) \triangleq \begin{bmatrix} ||Y_1(q(t),\dot{q}(t),\ddot{q}(t))|| \\ ||Y_2(q(t),\dot{q}(t),\ddot{q}(t))|| \end{bmatrix} \tag{3.123}$$

where $Y_i(\cdot)(i = 1,2)$ denotes $i^{th}$ row of the matrix $Y(\cdot)$ defined by (3.91). Here the maximal $\delta$, say $\hat{\delta}$, is the radius of the largest ball that fits inside the convex polytope $\Theta(\tau)$ and $\hat{\theta}$ is its center. That is, $\hat{\theta}$ is a point in $\Theta(\tau)$ that is as far as possible from $\delta\Theta(\tau)$ and $\hat{\delta}$ is the distance of $\hat{\theta}$ from $\delta\Theta(\tau)$.

Besides the batch-type approach linear programming (3.121), a recursive algorithm for (3.121) is also possible because the unfalsified controller parameter set $\Theta(\tau)$ is the intersection of degenerate ellipsoids (regions between 'parallel' hyperplanes), the recursive algorithm of minimal-volume outer approximation in [35] can be useful for the calculation of the intersections

In addition to the manipulator dynamics, the following first order transfer functions of different bandwidths are used to simulate the actuator dynamics:

$$G_a(s) = \frac{1}{\tau s + 1} \tag{3.124}$$

where $\frac{1}{2\pi\tau}$ is the actuator bandwidth in Hz; the values $\tau = 0$, $\tau = \frac{1}{10\pi}$ and $\tau = \frac{1}{40\pi}$ were used in the simulation and displayed in the plots. The parameters for the manipulator dynamics are chosen as:

$$m_1 = 1; \ ,l+1-1, \ ,m_e = 2, \delta_e = 30^o \tag{3.125}$$

$$I_1 = 0.12, l_{c_1} = 0.5, I_e = 0.25, l_{c_e} = 0.6 \tag{3.126}$$

so that the exact parameter vector (in the absence of actuator) is $\theta^* \triangleq [\theta_1, \theta_2, \theta_3, \theta_4]^T = [3.345, 0.97, 1.0392, 0.6]^T$. The scenario is as follows: the end effector mass $m_e$ changes back and forth between 2 and 20 periodically with period 0.5 $s$, and the inertia $I_e$ changes between 0.25 and 2.5, so that the parameter vector changes between [3.34, 0.97, 1.0392, 0.6] and [30.07, 9.7, 10.3923, 6] periodically with period 0.5 $s$ accordingly. The magnitudes of parameter vectors are unknown to the controller. The preferred trajectory used is

$$q_{d_1}(t) = 30^o(1 - \cos 2\pi t), \quad q_{d_2}(t) = 45^o(1 - \cos 2\pi t) \tag{3.127}$$

The external torque disturbance acting on the two joints are $\sin 20\pi t$ and $2\sin 13\pi t$, respectively. At time $t = 0$, the system is initially at rest with joint angles $q_1(0) = q_2(0) = 0.4rad$. The unfalsified control method is then compared with the adaptive method of [92]. The controller in [92] is:

$$u = Y_{slot}\hat{\theta} + K_D\dot{\tilde{q}} + \Lambda\tilde{q} \tag{3.128}$$

where $\hat{\theta}$ is the estimated parameter vector of $\theta$, and $K_D$ and $\Lambda$ are positive definite matrices. Also, $Y_{slot}$ satisfies

$$H(\theta, q)\ddot{q}_r + C(\theta, q, \dot{q})\dot{q}_r + g(\theta, q) = Y_{slot}(q, \dot{q}, \dot{q}_r, \ddot{q}_r)\theta \tag{3.129}$$

in which $\dot{q}_r = \dot{q}_d + \Lambda\tilde{q}$. The parameter update law for $\hat{\theta}$ is

$$\dot{\hat{\theta}} = -\Gamma Y_{slot}^T(\dot{\tilde{q}} + \Lambda\tilde{q}) \tag{3.130}$$

in which $\Gamma > 0$. The parameters used in the simulation are $K_D = 100I_2$, $\Lambda = 20I_2$, and $\Gamma = \mathrm{diag}([0.03,\ 0.05,\ 0.1,\ 0.3])$.

For the unfalsified control method $\Theta_0$ is taken as a solid square box centered at the origin with each edge of length 200. For simplicity, the computation delay time is taken as a constant value $10^{-3}$ $s$.

The parameter $\lambda$ used in the control law is $\lambda = 20$. The linear programming parameter update law is used, in which constant $\bar{d}(t) = [2, 4]^T$ is the bound on the effect of external disturbance in the performance specification $T_{spec}$. In this simulation, the 'correct' parameter vector changes periodically every 0.5 $s$ starting at $\tau = 0$ and at these times the parameter update law (3.117) is reset by discarding previous data, resetting the parameter update clock time to $\tau = 0$ and setting $\Theta(0) = \Theta_0$.

In the simulation, both control methods use $\hat{\theta}(0) = [0.5, 0.5, 0.5, 0.5]^T$ as initial guess for the parameter estimate, and the results are shown in Figure 3.78. Simulations in both cases were attempted for each of the three values of actuator time-constant $\tau = 0, \frac{1}{40\pi}, \frac{1}{10\pi}$. However, instabilities observed with the method in [92] were too severe to permit the simulation to be performed with actuator bandwidths smaller than 10 $Hz$ (*i.e.* $\tau \geq \frac{2\pi}{10}$). Thus five time histories appear in each plot in Figure 3.78, three for the unfalsified methods and two for the method in [92].

The four plots on the left of Figure 3.77 show the time histories for the joint angle tracking errors $\tilde{q}_1$ and $\tilde{q}_2$. The smallest amplitude error corresponds to the unfalsified control approach with infinite actuator bandwidth ($\tau = 0$). Next amplitudes are the unfalsified-control tracking errors for $\tau = \frac{1}{40\pi}$ and $\tau = \frac{1}{10\pi}$. Increasing amplitude tracking-errors are shown for the Slotine *et al.* controller with $\tau = 0$ and $\tau = \frac{1}{40\pi}$.

Also shown in Figure 3.78 are the actuator torques required for both the unfalsified controller and the Slotine *et al.* controller. The three smaller amplitude actuator signals shown correspond to the unfalsified controller, even though these control signals also produce smaller tracking errors as shown in the right side four plots in Figure 3.77. The plots show that the unfalsified controller is able to achieve a more precise and 'sure-footed' control over the arm's response without any appreciable increase in control energy. Figure 3.78 shows the estimated parameters $\hat{\theta}_i(t)$. The two sluggishly smooth traces in each of the four plots are for the controller in [92]. The crisp 'square-wave' response shown in the four plots are for the unfalsified controller. There is no perceptible difference in unfalsified controller response due to variations in the actuator time constant; the three chosen values of $\tau$ produce the

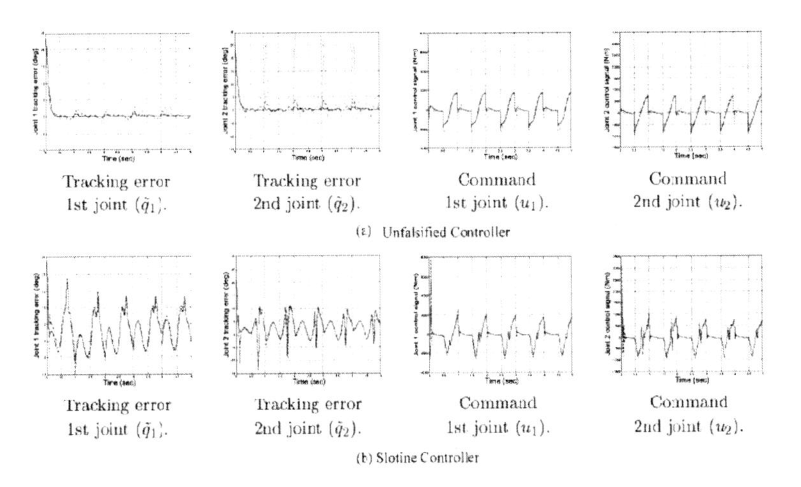


**Fig. 3.77** Comparison of tracking errors $\tilde{q}$ and control signals $u$ indicates that the unfalsified controller produces a quicker, more precise response, without increased control effort (solid line: ideal actuator, dashed line: 20 $Hz$ actuator dynamics, dotted line: 5 $Hz$ actuator dynamics).

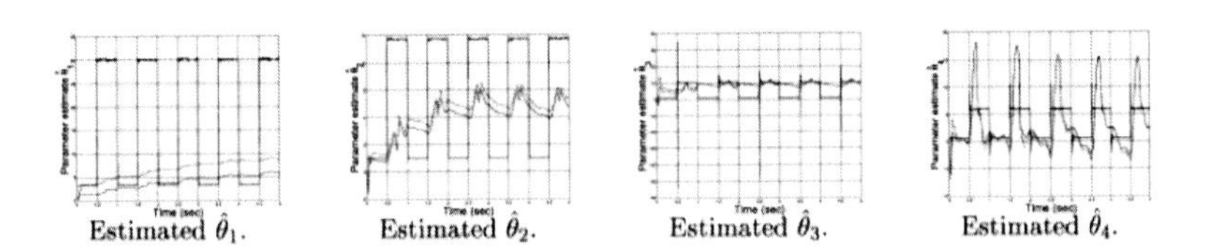


**Fig. 3.78** Simulation results for unfalsified controller (thick solid line: ideal actuator, thick dashed line: 20 $Hz$ actuator dynamics, thick dotted line: 5 $Hz$ actuator dynamics) and Slotine's controller (thin solid line: ideal actuator, thin dashed line: 20 $Hz$ actuator dynamics).

time histories that coincide with the 'square-wave' plots shown in the figure. The simulations show that the Slotine *et al.* controller cannot accurately track the 'correct' parameters. Attempts in [99] to improve this situation by adjusting Slotine's parameter $\Gamma$ proved unsuccessful.

Finally, the number of floating point operations (flops) required for each update of $\hat{\theta}(\tau)$ by the unfalsified controller's linear programming routine (solved by using the MATLAB Optimization Toolbox function `lp.m`) is plotted in Figure 3.79. The figure shows the times $\tau$ at which the controller gain $\hat{\theta}(\tau)$ was falsified and the number of flops (floating point operations) required to solve the linear program 3.121 to 3.122 to compute a new, as yet unfalsified controller at each of these times. As the figure shows, between 6 and 18 such falsifications occurred in each 0.5 $s$ interval between controller resets. The average computational load during the 5 $s$ simulation was about 0.8 $kflops/ms$. Though the computational burden of the unfalsified controller may seem large, it is well within the capacity of standard microprocessors.

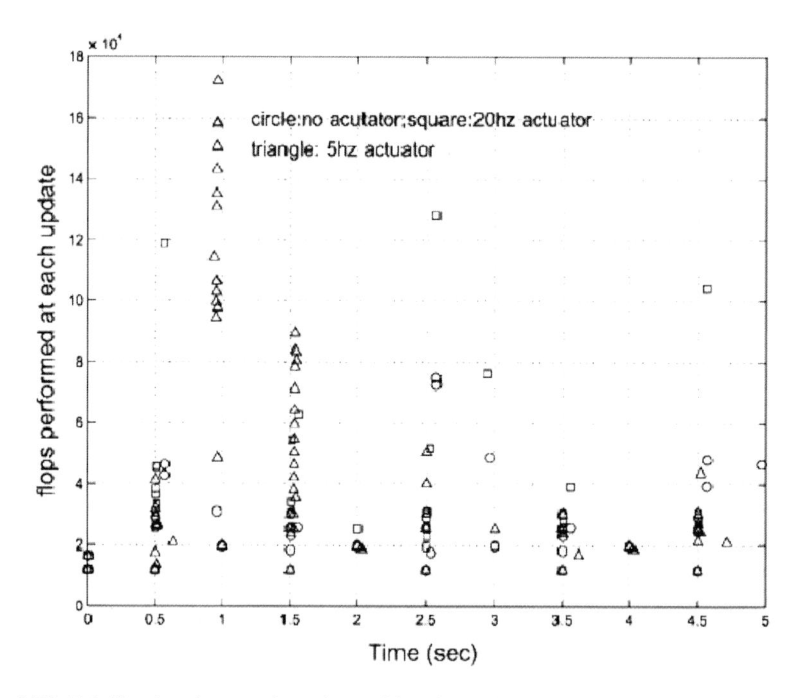


**Fig. 3.79** Falsification times and numbers of floating point operations (flops) at each of these times by linear program parameter update law

A recent publication [81] evaluates and confirms the above results.

# Chapter 4
# Conclusion

Motivated by the heightened interest in adaptive learning control systems and a proliferation of results in nonlinear stability theory, the authors studied and presented the *safe switching adaptive control* of the highly uncertain systems. Those are the systems whose dynamics, parameters, or uncertainty models may be insufficiently known for a variety of reasons, including, for example, difficulty in obtaining accurate model estimates, changes in subsystem dynamics, component failures, external disturbances, time variation of plant parameters (slow but consistent time variation or infrequent large jumps) *etc*. Control of uncertain systems has traditionally been attempted using, on the one hand, robust control techniques (classical $H_\infty$ robust control and modern enhancements using linear matrix inequalities (LMI) conditions and integral quadratic constraints (IQC)) [25], [91], [120], whose proofs of robust stability and performance hinge upon the knowledge of sufficiently small uncertainty bounds around the nominal model. On the other hand, adaptive control techniques aim to further enhance robustness for larger uncertainties by introducing an outer adaptive loop that adjusts (tunes) controller parameters based on the observed data. Both streams have inherent limitations: robust control methods are valid insofar as the proposed models match the actual plant and uncertainty bounds; and adaptation in conventional continuous adaptive tuning may be slow in comparison to the swiftly changing plant dynamics/parameters or rapidly evolving environment, thus yielding unacceptable performance or even instability in the practical terms. The fact that a single controller (fixed or adaptive) may not be able to cope with insufficiently known or changing plant was the primary reason that brought forth the notion of switching in the context of adaptive control. A wide variety of switching algorithms have been proposed in the last 20 years, with nearly all of them basing their stability/performance proofs to some extent on prior assumptions, which are invariably difficult to verify and may fail to hold in practice.

This mismatch between the reality and prior assumptions is the core of the problem which has been addressed in this book. In the preceding pages, a theoretical explanation of, and a solution to, the model mismatch stability problem associated with a majority of adaptive control design techniques has been given. The algorithm and the methodology of adaptive switching control proposed in this monograph

M. Stefanovic and M.G. Safonov: Safe Adaptive Control, LNCIS 405, pp. 129–130.
springerlink.com 

are based on the theory of control law unfalsification according to which a reliable adaptive control law is synthesized.

Adaptive and learning control techniques have a significant potential to enhance robustness of stability and performance of the systems operating under uncertain conditions. For example, adverse operating conditions, to which aircraft control systems are often subjected, introduce impacts and risks that are difficult to anticipate, calling for a reliable and prompt control action. The result proposed here belongs to the class of control paradigms that fully utilize information in the accumulated experimental data, and maximize robustness by introducing as few prior assumptions as are presently known, while at the same time converging quickly to a stabilizing solution, often within a fraction of an unstable plants largest unstable time constant. Thus, it forms a particularly attractive solution for the design and analysis of the fast adaptive fail-safe recovery systems for battle-damaged aircraft control systems, missile guidance systems, reconfigurable communication networks, precision pointing and tracking systems.

Some directions for future research in the following are listed. An important subject is the application of the proposed safe adaptive control paradigm in a variety of real life scenarios. Computational solvability of the algorithm (*e.g.*, polynomial-time type) needs to be investigated, particularly for the case when the set of candidate controllers is continuously parameterized. Tractability issues may depend to some extent on the compactness of the candidate controller set, and on its representability as a finite union of convex sets. Tools from the theoretical computer science and artificial intelligence concepts (such as machine learning [70]) will be used to characterize and enhance levels of algorithm solvability.

On the theoretical side, it is of interest to further explore efficient ways to continuously and adaptively generate new candidate controllers on the fly, enhancing the system with an additional supervisory loop with a hypothesis generating role. The theory presented in this book relies on the sole assumption that the adaptive control problem, posed as optimization problem, is feasible, which means that the solution exists in the pool of candidate controllers. To the best of our knowledge, this assumption underlies, implicitly or explicitly, all other adaptive schemes so that it is minimal. If it happens, however, that this assumption does not hold (*e.g.*, when one starts out with an initially sparse set of controllers), then it is needed to have a certain hypothesis generator that will create new candidate controllers as the system evolves.

# Appendix A
# Relation Between $||\tilde{r}_{K_N}||_t$ and $||r||_t$, $\forall t \geq 0$

**Lemma A.1.** *Consider the switching feedback adaptive control system $\Sigma$ (Figure 2.1), where uniformly bounded reference input $r$, as well as the output $z = [u, y]$ are given. Suppose there are finitely many switches. Let $t_N$ and $K_N$ denote the final switching instant and the final switched controller, respectively. Suppose that the final controller $K_N$ is SCLI (*i.e.*, the fictitious reference signal $\tilde{r}_{K_N}(z,t)$ is unique and incrementally stable). Then,*

$$||\tilde{r}_{K_N}||_t < ||r||_t + \alpha < \infty, \; \forall t \geq 0 \,. \tag{A.1}$$

*Proof.* By the assumption there are finitely many switches. Consider the control configuration in Figure A.1. The top branch generates the fictitious reference signal of the controller $K_N$. Its inputs are the measured data $(y, u)$, and its output is $\tilde{r}_{K_N}$. The output is generated by the fictitious reference signal generator for the controller $K_N$, denoted by $\mathfrak{K}^N_{CLI}$. In the middle interconnection, the signal $u_N$, generated as the output of the final controller $K_N$ excited by the actual applied signals $r$ and $y$, is simply inverted by passing through the causal left inverse $\mathfrak{K}^N_{CLI}$. Finally, the bottom interconnection has the identical structure as the top interconnection (series connection of $\hat{K}_t$ and $\mathfrak{K}^N_{CLI}$), except that it should generate the actually applied reference signal $r$. To this end, another input to the bottom interconnection is added (denoted $\omega$), as shown in Figure A.1. This additional input $\omega$ can be thought of as a compensating (bias) signal, that accounts for the difference between the subsystems generating $r$ and $\tilde{r}_{K_N}$ before the time of the last switch. In particular, it can be shown (as seen in Figure A.1) that $\omega \triangleq P_{t_N}(u_N - u)$ (due to the fact that $u_N(t) \equiv u(t)$, $\forall t \geq t_N$).

By definition, $\mathfrak{K}^N_{CLI}$ is incrementally stable. Thus, according to Definition 2.2, there exist constants $\tilde{\beta}, \tilde{\alpha} \geq 0$ such that

$$\begin{aligned} ||\tilde{r}_{K_N} - r||_t &\leq \tilde{\beta} \cdot ||u - u_N||_t + \tilde{\alpha} \\ &\leq \tilde{\beta} \cdot ||\omega||_{t_N} + \tilde{\alpha} \\ &< \infty \; \forall t \geq 0 \,. \end{aligned} \tag{A.2}$$

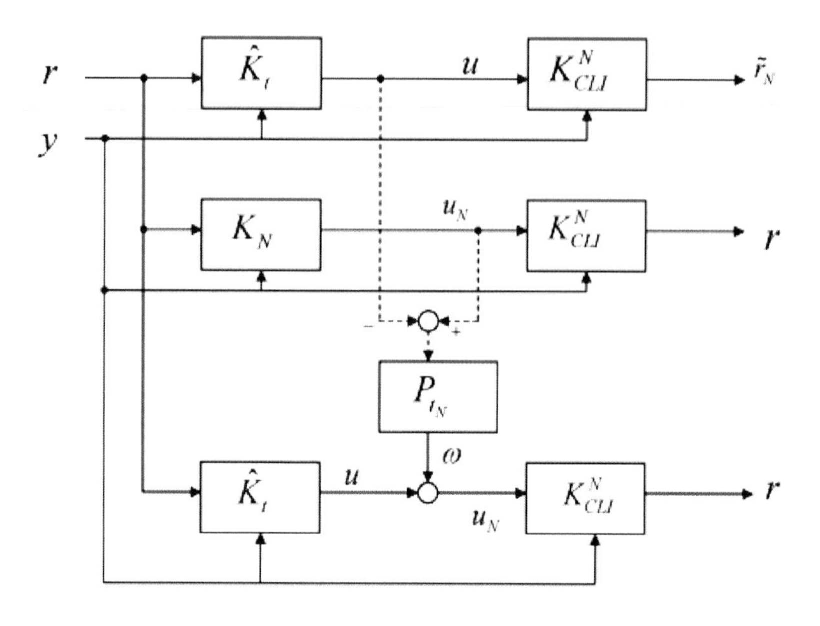


**Fig. A.1** Generators of the true and fictitious reference signals

Hence by the triangle inequality for norms, inequality (A.1) holds with

$$\alpha = \tilde{\beta} \cdot \|\omega\|_{t_N} + \tilde{\alpha} \ . \tag{A.3}$$

# Appendix B
# Matrix Fraction Description of $\tilde{r}$

**Definition B.1** [108]. The ordered pair $(N_K, D_K)$ is a *left matrix fraction description* (MFD) of a controller $K$ if $N_K$ and $D_K$ are stable, $D_K$ is invertible, and $K = D_K^{-1} N_K$.

*Remark Appendix B.1* To avoid restricting our attention to only those controllers that are *stably* causally left invertible, *i.e.*, controllers whose fictitious reference signal (FRS) generator is stable, we can use the MFD representation of the controllers and write a *modified fictitious reference* signal as

$$\tilde{v}_K = D_K u + (-N_K)(-y) \tag{B.1}$$

and

$$\tilde{v}_K = N_K \tilde{r}_K$$

as in [65]. Similarly, $v = N_{\hat{K}_t} r$ would represent the modified applied reference signal $r$, which is related to the active controller in the loop $\hat{K}_t$. Thus, although $\tilde{r}_K$ may not be stable, which is the case with the non-minimum phase controllers, $\tilde{v}_K$ is, by construction.

# Appendix C
# Unfalsified Stability Verification

Stability verification for the cost function (2.7) in Section 2.4 is performed below. Recall that the controller switched in the loop at time $t_i$ is denoted $K_i$, and $i \in I \doteq \{0,1,...,N\}$, $N \in \mathbb{N} \cup \{\infty\}$ are the indices of the switching instants. When $t_0 = 0$, we have $\forall K \in \mathbf{K}$, $V(K,z,t_0) = \beta + \gamma||K||^2 > 0$. Let the controller switched at time $t_i$ be denoted $K_i$ (thus, $\hat{K}_{t_i} = K_i$). Then, owing to the cost minimization property of the switching algorithm, $K_0 = \arg\min_K V(K,z,0)$, and $V(K_0,z,0) = \beta + \gamma||K_0||^2 > 0$.

Denote by $t = t_N$ the time of the final switch, and the corresponding controller $K_N$. Consider the time interval $[0,t_1)$. During this time period, the active controller in the loop is $\hat{K}_t = K_0$.

$$\begin{aligned} V(K_0,z,t_1^-) &= V(K_0,z,t_1) \\ &= \max_{\tau \le t_1} \frac{||y||_\tau^2 + ||u||_\tau^2}{||\tilde{r}_{K_0}||_\tau^2 + \alpha} + \beta + \gamma||K_0||^2 \\ &= \max_{\tau \le t_1} \frac{||y||_\tau^2 + ||u||_\tau^2}{||r||_\tau^2 + \alpha} + \beta + \gamma||K_0||^2, \end{aligned}$$

since $\tilde{r}_{K_0} \triangleq \tilde{r}(K_0,z,t) \equiv r(t)$, $t \in [0,t_1)$.

Since $r$ is uniformly bounded, $||r||_t^2 = \int_0^t |r(\tau)|^2 d\tau < \infty$.

At $t = t_1$, the cost of the current controller exceeds the current minimum by $\varepsilon$:

$$\begin{aligned} V(K_0,z,t_1) &= \max_{\tau \le t_1} \frac{||y||_\tau^2 + ||u||_\tau^2}{||\tilde{r}_{K_0}||_\tau^2 + \alpha} + \beta + \gamma||K_0||^2 \\ &= \varepsilon + \min_K V(K,z,t_1), \end{aligned} \tag{C.1}$$

and so, according to the hysteresis switching algorithm, a switch to the controller $K_1 \doteq \arg\min_K V(K,z,t_1)$ takes place. Expression in (C.1) is finite since $\varepsilon$ is finite and

$$\min_K V(K,z,t_1) \le \sup_{t \in \mathbf{T}, z \in \mathbf{Z}} \min_K V(K,z,t) \triangleq V_{true}(K_{RSP}),$$

where $V_{true}(K_{RSP})$ is finite because of the feasibility assumption. Denoting the sum $\varepsilon + \min_K V(K,z,t_1)$ by $\psi_1$, we have:

$$
\begin{aligned}
&\max_{\tau \le t_1} \frac{||y||_\tau^2 + ||u||_\tau^2}{||\tilde{r}_{K_0}||_\tau^2 + \alpha} + \beta + \gamma ||K_0||^2 = \psi_1, \\
&\Rightarrow ||y||_{t_1}^2 + ||u||_{t_1}^2 + (\beta + \gamma ||K_0||^2)(||\tilde{r}_{K_0}||_{t_1}^2 + \alpha) \\
&\qquad \le \ \psi_1 (||\tilde{r}_{K_0}||_{t_1}^2 + \alpha), \\
&\Rightarrow ||y||_{t_1}^2 + ||u||_{t_1}^2 + (\beta + \gamma ||K_0||^2)(||r||_{t_1}^2 + \alpha) \\
&\qquad \le \psi_1 (||r||_{t_1}^2 + \alpha) < \infty, \\
&\Rightarrow ||y||_{t_1} < \infty, \ \ ||u||_{t_1} < \infty \, .
\end{aligned}
$$

Now consider the next switching period, $[t_1, t_2)$. The active controller in the loop is $\hat{K}(t) = K_1$. Hence,

$$
\begin{aligned}
0 < \beta + \gamma ||K_1||^2 \le V(K_1, z, t_1) &\triangleq \min_K V(K, z, t_1) \\
&= \max_{\tau \le t_1} \frac{||y||_\tau^2 + ||u||_\tau^2}{||\tilde{r}_{K_1}||_\tau^2 + \alpha} + \beta + \gamma ||K_1||^2 < \infty
\end{aligned}
$$

where the second inequality from the left follows from the monotone increasing property of $V$.

Therefore, $||\tilde{r}_{K_1}||_{t_1} < \infty$. Next,

$$
\begin{aligned}
\infty > \varepsilon + \min_K V(K, z, t_2) &= V(K_1, z, t_2) \\
\triangleq \max_{\tau \le t_2} \frac{||y||_\tau^2 + ||u||_\tau^2}{||\tilde{r}_{K_1}||_\tau^2 + \alpha} &+ \beta + \gamma ||K_1||^2 \\
&\ge \beta + \gamma ||K_1||^2 > 0 \, .
\end{aligned}
$$

Thus, $||\tilde{r}_{K_1}||_{t_2}$ is finite, and so are $||y||_{t_2}$, $||u||_{t_2}$. By induction, we conclude that

$$
||y||_{t_N} \le \infty, \ \ ||u||_{t_N} \le \infty \tag{C.2}
$$

where $t_N$ is the final switching time. Since $t_N$ is the final switching time:

$$
\begin{aligned}
0 < \beta + \gamma ||K_N||^2 \le V(K_N, z, t) = \\
\max_{\tau \le t} \frac{||y||_\tau^2 + ||u||_\tau^2}{||\tilde{r}_{K_N}||_\tau^2 + \alpha} + \beta + \gamma ||K_N||^2 \\
< \varepsilon + \min_K V(K, z, t), \quad \forall t \ge t_N \, .
\end{aligned}
$$

Thus, $||\tilde{r}_{K_N}||_t$ is finite for any finite $t > t_N$. Further,

$$
\begin{aligned}
0 < \beta + ||K_N||^2 &\leq \sup_{t \in \mathbb{R}_+} V(K_N, z, t) \\
&= \sup_{t \in \mathbb{R}_+} \max_{\tau \leq t} \frac{||y||_\tau^2 + ||u||_\tau^2}{||\tilde{r}_{K_N}||_\tau^2 + \alpha} + \beta + \gamma ||K_N||^2 \\
< \varepsilon + \sup_{t \in \mathbb{R}_+} \min_K V(K, z, t) &\leq \varepsilon + V_{true}(K_{RSP}) < \infty \quad \forall t, \\
&\Rightarrow || \begin{bmatrix} y \\ u \end{bmatrix} ||_t < \beta_2 ||\tilde{r}_{K_N}||_t + \alpha_2 \ \ \forall t > 0, \\
&\quad \text{for some } \beta_2, \alpha_2 \geq 0 .
\end{aligned}
$$

From the above it is concluded that the stability of the closed loop switched system with the final controller $K_N$ is unfalsified by $(\tilde{r}_{K_N}, z)$.

# References


[1] Agrawal, D., Granelli, F.: Redesigning an active queue management system. IEEE Globecom 2, 702–706 (2004)

[2] Anderson, B.D.O.: Adaptive systems, lack of persistency of excitation and bursting phenomena. Automatica 21, 247–258 (1985)

[3] Anderson, B.D.O.: Failures of adaptive control theory and their resolution. Communications in Information and Systems 5(1), 1–20 (2005)

[4] Anderson, B.D.O., Brinsmead, T.S., Liberzon, D., Morse, A.S.: Multiple model adaptive control with safe switching. Int. J. Adaptive Control and Signal Processing 15, 445–470 (2001)

[5] Arehart, A.B., Wolovich, W.A.: Bumpless switching controllers. In: Proc. IEEE Conference on Decision and Control, pp. 1654–1655 (December 1996)

[6] Åström, K.J.: Theory and applications of adaptive control - a survey. Automatica 195, 471–486 (1983)

[7] Åström, K.J., Wittenmark, B.: Adaptive Control. Addison Wesley, Reading (1995)

[8] Fekri Asl, S., Athans, M., Pascoal, A.: Issues, progress and new results in robust adaptive control. Int. J. Adaptive Control and Signal Processing 20, 519–579 (2006)

[9] Balakrishnan, J.: Control System Design Using Multiple Models, Switching and Tuning. Ph.D. Thesis, Yale University, New Haven, CT (1996)

[10] Balakrishnan, J., Narendra, K.S.: Improving transient response of adaptive control systems using multiple models and switching. IEEE Trans. Autom. Control 39, 1861–1866 (1994)

[11] Balakrishnan, J., Narendra, K.S.: Adaptive control using multiple models. IEEE Trans. Autom. Control 42 (1997)

[12] Baldi, S., Battistelli, G., Mosca, E., Tesi, P.: Multi-model unfalsified adaptive switching supervisory control. Automatica 46(2), 249–259 (2010)

[13] Battistelli, G., Mosca, E., Safonov, M.G., Tesi, P.: Stability of unfalsified adaptive switching control in noisy environments. IEEE Trans. Autom. Control (to appear, 2010)

[14] Bertsekas, D.P.: Nonlinear Programming, 2nd edn. Athena Scientific, Belmont (1999)

[15] Brozenec, T.F., Tsao, T.C., Safonov, M.G.: Convergence of control performance by unfalsification of models – levels of confidence. Int. J. of Adaptive Control and Signal Processing 15(5), 431–444 (2001)

[16] Brugarolas, P., Fromion, V., Safonov, M.G.: Robust Switching Missile Autopliot. In: Proc. 1998 American Control Conference, Philadelhia, PA, USA (1998)

[17] Brugarolas, P., Safonov, M.G.: A canonical representation for unfalsified control in truncated spaces. In: Proc. 1999 IEEE International Symposium on Computer Aided Control System Design, Hawaii, USA (1999)
[18] Cabral, F.B., Safonov, M.G.: Unfalsified model reference adaptive control using the ellipsoid algorithm. Int. J. Adaptive Control and Signal Processing 18(8), 605–714 (2004)
[19] Campi, M.C., Lecchini, A., Savaresi, S.M.: Virtual reference feedback tuning: A direct method for the design of feedback controllers. Automatica 38, 1337–1346 (2002)
[20] Cao, J., Stefanovic, M.: Performance improvement in unfalsified control using neural networks. In: Proc. 17th World Congress of the International Federation of Automatic Control, Seoul, Republic of Korea, pp. 6536–6541 (2008)
[21] Cao, J., Stefanovic, M.: Switching congestion control for satellite TCP/AQM networks. In: Proc. American Contr. Conf., St. Louis, MO (June 2009)
[22] Chang, M.W., Safonov, M.G.: Unfalsified adaptive control: the benefit of bandpass filters. In: AIAA Guidance, Navigation and Control Conf. and Exhibit, Honolulu, HI (2008)
[23] Cheong, S.-Y., Safonov, M.G.: Improved bumpless transfer with slow-fast controller decomposition. In: Proceedings of the American Control Conference. St. Louis, MO, pp. 4346–4350
[24] Chiang, M.L., Hua, C.C., Lin, J.R.: Direct Power Control for Distributed PV Power System. In: Power Conversion Conference (PCC), Osaka, Japan (2002)
[25] Chiang, R.Y., Safonov, M.G.: Robust Control Toolbox. Mathworks, South Natick (1988)
[26] Dehghani, A., Anderson, B.D.O., Lanzon, A.: Unfalsified adaptive control: A new controller implementation and some remarks. In: European Control Conference ECC 2007, Kos, Greece (July 2007)
[27] Dehghani, A., Lecchini-Visintini, A., Lanzon, A., Anderson, B.D.O.: Validating controllers for internal stability utilizing closed-loop data. IEEE Trans. Autom. Control 54(11), 2719–2725 (2009)
[28] Drenick, R.F., Shahbender, R.A.: Adaptive servomechanisms. AIEE transactions 76, 286–292 (1957)
[29] Egardt, B.: Global stability analysis of adaptive control systems with disturbances. In: Proc. Joint Automatic Control Conference, San Fransisco, CA (1980)
[30] Engell, S., Tometzki, T., Wonghong, T.: A new approach to adaptive unfalsified control. In: Proc. European Control Conf., Kos, Greece, pp. 1328–1333 (July 2007)
[31] Feldbaum, A.A.: Optimal Control Systems. Academic Press, New York (1965)
[32] Feng, W.-C., Kandlur, D.D., Saha, D., Shin, K.G.: A Self-Configuring RED Gateway. In: Proceedings of INFOCOM 1999, Eighteenth Annual Joint Conference of the IEEE Computer and Communications Societies (1999)
[33] Floyd, S., Gummadi, R., Shenker, S.: Adaptive RED: An Algorithm for Increasing the Robustness of RED's Active Queue Management (2001), `http://www.icir.org/floyd/papers/adaptiveRed.pdf`
[34] Floyd, S., Jacobson, V.: Random Early Detection gateways for congestion avoidance. IEEE/ACM Trans. on Networking 1, 397–413 (1993)
[35] Fogel, E., Huang, Y.F.: On the value of information in system identification-bounded noise case. Automatica 18(2), 229–238 (1982)
[36] Freeman, S., Herron, J.C.: Evolutionary Analysis, 4th edn. Benjamin Cummings publishing (2007)
[37] Fu, M., Barmish, B.R.: Adaptive stabilization of linear systems via switching control. IEEE Trans. Autom. Control 31(12), 1097–1103 (1986)

[38] Goodwin, G.C., Mayne, D.Q.: A parameter estimation perspective of continuous time adaptive control. Automatica 23 (1987)
[39] Graebe, S.F., Ahlén, A.L.B.: Dynamic transfer among alternative controllers and its relation to antiwindup controller design. IEEE Trans. Control Systems Techcology 4(1), 92–99 (1996)
[40] Hanus, R., Kinnaert, M., Henrotte, J.-L.: Conditioning technique, a general antiwindup and bumpless transfer method. Automatica 23(6), 729–739 (1987)
[41] Hespanha, J.P.: Logic-Based Switching Algorithms in Control. Ph.D. Thesis, Yale University, New Haven, CT (1998)
[42] Hespanha, J.P., Morse, A.S.: Certainty equivalence implies detectability. Systems and Control Letters 1(1), 1–13 (1999)
[43] Hespanha, J.P., Morse, A.S.: Stability of switched systems with average dwell time. In: Proc. 38th Conf. on Decision and Control (1999)
[44] Hespanha, J., Liberzon, D., Morse, A.S., Anderson, B.D., Brinsmead, T., De Bruyne, F.: Multiple model adaptive control, Part 2: switching. Int. J. Robust and Nonlinear Control 11, 479–496 (2001)
[45] Hespanha, J.P., Liberzon, D., Morse, A.S.: Supervision of integral-input-to-state stabilizing controllers. Automatica 38(8), 1327–1335 (2002)
[46] Hespanha, J.P., Liberzon, D., Morse, A.S.: Hysteresis-based switching algorithms for supervisory control of uncertain systems. Automatica 39(2) (2003)
[47] Hespanha, J.P., Liberzon, D., Morse, A.S.: Overcoming the limitations of adaptive control by means of logic-based switching. Systems and Control Letters 49, 49–65 (2003)
[48] Hollot, C.V., Misra, V., Towsley, D., Gong, W.: Analysis and design of controllers for AQM routers supporting TCP flows. IEEE Transactions on Automatic Control 47, 945–959 (2002)
[49] Hsieh, G.C., Safonov, M.G.: Conservatism of the gap metric. IEEE Trans. Autom. Control 38(4), 594–598 (1993)
[50] Ingimundarson, A., Sanchez-Pena, R.S.: Using the unfalsified control concept to achieve fault tolerance. In: IFAC World Congress, Seoul, Korea, pp. 1236–1242 (July 2008)
[51] Ioannou, P.A., Datta, A.: Robust adaptive control: A unified Approach. Proc. of the IEEE 79(12), 1735–1768 (1991)
[52] Ioannou, P.A., Sun, J.: Theory and design of robust direct and indirect adaptive control schemes. Int. J. Control 47(3), 775–813 (1988)
[53] Ioannou, P.A., Sun, J.: Stable and Robust Adaptive Control. Prentice Hall, Englewood-Cliffs (1996)
[54] Jun, M., Safonov, M.G.: Automatic PID tuning: an application of unfalsified control. Proc. IEEE CCA/CACSD 2, 328–333 (1999)
[55] Kaneko, O.: On linear canonical controllers within the unfalsified control framework. In: IFAC World Congress, Seoul, Korea, pp. 12279–12284 (July 2008)
[56] Kim, K.B.: Design of feedback controls supporting TCP based on the state space approach. IEEE Trans. Autom. Control 51(7) (2006)
[57] Kosmatopoulos, E., Ioannou, P.A.: Robust switching adaptive control of multi-input nonlinear systems. IEEE Trans. Autom. Control 47(4), 610–624 (2002)
[58] Kosut, R.L.: Uncertainty model unfalsification for robust adaptive control. In: Proc. IFAC Workshop on Adaptive Systems in Control and Signal PRocessing, Glasgow, Scotland, UK (1998)
[59] Kokotovic, P.V., Krstic, M., Kanellakopoulos, I.: Backstepping to passivity: recursive design of adaptive systems. In: Proc. IEEE Conf. on Decision and Control, Tucson, AZ, pp. 3276–3280 (1992)

[60] Kreisselmeier, G., Anderson, B.D.O.: Robust model reference adaptive control. IEEE Trans. Autom. Control 31(2), 127–133 (1986)

[61] Kulkarni, S.R., Ramadge, P.J.: Model and controller selection policies based on output prediction errors. IEEE Trans. Autom. Control 41, 1594–1604 (1996)

[62] Landau, I., Lozano, R., M'Saad, M.: Adaptive Control. In: Communications and Control Engineering series. Springer, Heidelberg (1998)

[63] Low, S., Paganini, F., Wang, Z., Doyle, J.: A new TCP/AQM for stable operation in fast networks. In: Proceedings of IEEE INFOCOM (2003)

[64] Manfredi, S., di Bernardo, M., Garofalo, F.: Robust output feedback active queue management control in TCP networks. In: Proceedings of IEEE Conference on Decision and Control (2004)

[65] Manuelli, C., Cheong, S.G., Mosca, E., Safonov, M.G.: Stability of Unfalsified adaptive control with non SCLI controllers and related performance under different prior knowledge. In: European Control Conference ECC 2007, Kos, Greece (2007)

[66] Markovsky, I., Rapisarda, P.: On the linear quadratic data-driven control. In: European Control Conference ECC 2007, Kos, Greece, pp. 5313–5318 (2007)

[67] Mårtensson, B.: The order of any stabilizing regulator is sufficient information for adaptive stabilization. Systems and Control Letters 6(2), 87–91 (1985)

[68] Mårtensson, B., Polderman, J.W.: Correction and simplification to The order of any stabilizing regulator is sufficient a priori information for adaptive stabilization. Systems and Control Letters 20(6), 465–470 (1993)

[69] Misra, V., Gong, W., Towsley, D.: Fluid-based analysis of network of AQM routers supporting TCP flows with an application to RED. In: Proc. of ACM/SIGCOMM 2000 (2000)

[70] Mitchell, T.: Machine Learning. McGraw Hill, New York (1997)

[71] Morse, A.S.: Global stability of parameter adaptive control systems. IEEE Trans. Autom. Control, 433–439 (1980)

[72] Morse, A.S.: Supervisory control of families of linear set-point controllers — Part I: exact matching. IEEE Trans. Autom. Control 41(10), 1413–1431 (1996)

[73] Morse, A.S.: Supervisory control of families of linear set-point controllers — part 2: robustness. IEEE Trans. Autom. Control 42, 1500–1515 (1997)

[74] Morse, A.S.: Lecture notes on logically switched dynamical systems. In: Nistri, P., Stefani, G. (eds.) Nonlinear and Optimal control theory lectures, C.I.M.E. summer school, Cetraro, Italy. Springer, Heidelberg (2008)

[75] Morse, A.S., Mayne, D.Q., Goodwin, G.C.: Applications of hysteresis switching in parameter adaptive control. IEEE Trans. Autom. Control 37(9), 1343–1354 (1992)

[76] Narendra, K.S., Annaswamy, A.M.: Stable Adaptive Systems. Prentice Hall, Inc., New Jersey (1988)

[77] Narendra, K.S., Balakrishnan, J.: Adaptive control using multiple models. IEEE Trans. Autom. Control 42(2), 171–187 (1997)

[78] Narendra, K.S., Taylor, I.H.: Frequency Domain Criteria for Absolute Stability. Academic Press, New York (1973)

[79] O'Dwyer, A.: Handbook of PI And PID Controller Tuning Rules. Imperial College Press, London (2006)

[80] Parks, P.C.: Lyapunov redesign of model reference adaptive control systems. IEEE Trans. Autom. Control 11, 362–367 (1966)

[81] Pawluk, M., Arent, K.: Unfalsified control of manipulators: Simulation analysis. Bulletin of the Polish Academy of Sciences Technical Sciences 53(1), 19–29 (2005)

[82] Popper, K.R.: The Logic of Scientific Discovery, 5th edn. Routledge, NewYork (2002)

[83] Praly, L.: Robust model reference adaptive controller (Part I: stability analysis). In: Proc. IEEE Conf. on Decision and Control (1984)

[84] Reichert, R.T.: Dynamic scheduling of modern robust control autopilot designs for missiles. IEEE Control Systems Magazine, 35–42 (1992)

[85] Rohrs, C.E., Valavani, L., Athans, M., Stein, G.: Robustness of adaptive control algorithms in the presence of unmodeled dynamics. IEEE Trans. Autom. Control 30, 259–294 (1985)

[86] Rudin, W.: Principles of Mathematical Analysis, 3rd edn. McGraw-Hill, Inc., New York (1976)

[87] Saeki, M., Hamada, O., Wada, N., Masubuchi, I.: PID gain tuning based on falsification using bandpass filters. In: SICE–ICASE Int. Joint Conf., Pusan, Korea (2006)

[88] Safonov, M.G.: Stability and Robustness in Multivariable Feedback Systems. MIT Press in Signal Processing, Optimization and Control (1980)

[89] Safonov, M.G., Jonckheere, E.A., Verma, M., Limebeer, D.J.N.: Synthesis of positive real multivariable feedback systems. Int. J. Control 45, 817–842 (1987)

[90] Safonov, M.G., Tsao, T.: The unfalsified control concept and learning. IEEE Trans. Autom. Control 42(6), 843–847 (1997)

[91] Skogestad, S., Postlethwaite, I.: Multivariable Feedback Control. John Wiley, New York (1996)

[92] Slotine, J.J.E., Li, W.: Adaptive manipulator control: a case study. IEEE Transactions on Automatic Control 33(11), 995–1003 (1988)

[93] Soma, S., Kaneko, O., Fujii, T.: Unfalsified control for real-time tuning of PID parameters. In: Proc. SICE 2003 Annual Conference, Fukui, Japan (2003)

[94] Stefanovic, M., Safonov, M.G.: Safe adaptive switching control: Stability and convergence. IEEE Trans. Autom. Control 53, 2012–2021 (2008)

[95] Stefanovic, M., Wang, R., Safonov, M.G.: Stability and convergence in Adaptive Systems. In: Proc. American Control Conference, Boston, MA (2004)

[96] Steinbuch, M., Van Helvoort, J., Aangenent, W., De Jager, B., Van de Molengraft, R.: Data-based control of motion systems. In: Proc. IEEE Conference on Control Applications, Toronto, Canada (2005)

[97] Sutton, R.S., Barto, A.G.: Reinforcement Learning: An Introduction. MIT Press, Cambridge (1998)

[98] Tanenbaum, A.S.: Computer Networks, 4th edn. Pearson Education, London (2003)

[99] Tsao, T.C., Safonov, M.G.: Unfalsified direct adaptive control of a two-link robot arm. Int. J. Adaptive Control and Signal Processing 15(3), 319–334 (2001)

[100] Tsao, T.C., Brozenec, T., Safonov, M.G.: Unfalsified Adaptive Spacecraft Attitude Control. In: Proc. AIAA Guidance, Naviation and Control Conf., Austin, TX, August 11-14 (2003)

[101] Truxal, J.G.: Adaptive control. In: Proc. 2nd World Congress of International Federation on Automatic Control (IFAC), Basle, Switzerland (1963)

[102] Turner, M.C., Walker, D.J.: Linear quadratic bumpless transfer. Automatica 36, 1089–1101 (2000)

[103] van Helvoort, J., de Jager, B., Steinbuch, M.: Sufficient conditions for data-driven stability of ellipsoidal unfalsified control. In: Proc. IEEE Conf. on Decision and Control, San Diego, CA, pp. 453–458 (2006)

[104] van Helvoort, J.: Unfalsified control: Data-driven control design for performance improvement. PhD thesis, University of Eindhoven (2007)

[105] van Helvoort, J., de Jager, A.G., Steinbuch, M.: Direct Data-driven recursive controller unfalsification with analytic update. Automatica 43(12), 2034–2046 (2007)

[106] van Helvoort, J., de Jager, A.G., Steinbuch, M.: Data-driven controller unfalsification with analytic update applied to a motion system. IEEE Trans. on Control Systems Technology 16(6), 1207–1217 (2008)
[107] van Helvoort, J., de Jager, A.G., Steinbuch, M.: Data-driven multivariable controller design using ellipsoidal unfalsified control. Systems and Control Letters 57(9), 759–762 (2008)
[108] Vidyasagar, M.: Control System Synthesis: A Factorization Approach. The MIT Press, Cambridge (1988)
[109] Wang, R., Paul, A., Stefanovic, M., Safonov, M.G.: Cost-detectability and stability of adaptive control systems. Int. J. Robust and Nonlinear Control, Special Issue: Frequency-domain and matrix inequalities in systems and control theory (dedicated to the 80th birthday of V. A. Yakubovich) 17(5-6), 549–561 (2007)
[110] Wheeden, R.L., Zygmund, A.: Measure and Integral: An Introduction to Real Analysis. Marcel Dekker, New York (1977)
[111] Whitaker, H.P., Yamron, J., Kezer, A.: Design of model reference adaptive control systems for aircraft. Report R-164, Instrumentation Laboratory. MIT Press, Cambridge (1958)
[112] Willems, J.C.: Paradigms and puzzles in the theory of dynamical Systems. IEEE Trans. Autom. Control 36(3), 259–294 (1991)
[113] Wonghong, T., Engell, S.: Application of a new scheme for adaptive unfalsified control to a CSTR. In: IFAC World Congress, Seoul, South Korea, pp. 13247–13250 (July 2008)
[114] Yakubovich, V.A.: On the Theory of Adaptive Systems. Dokl. Akad. Nauk SSSR 182(3), 518–522 (1968)
[115] Yame, J.J.: Reconfigurable control via online closed-loop system performance assessment. In: AIAA Guidance, navigation and Control Conference, AIAA 2005-6137, San Francisco, CA (2005)
[116] Zaccarian, L., Teel, A.R.: The L2 (l2) bumpless transfer problem: its definition and solution. In: Proc. IEEE Conference on Decision and Control, pp. 5505–5510 (December 2004)
[117] Zames, G.: On the input-output stability of nonlinear time-varying feedback systems, Part 1: conditions derived using concepts of loop gain, conicity and positivity. IEEE Trans. Autom. Control 11 (1966)
[118] Zhivoglyadov, P., Middleton, R.H., Fu, M.: Localization based switching adaptive control for time-varying discrete-time systems. IEEE Trans. Autom. Control 45(4), 52–75 (2000)
[119] Zhivoglyadov, P., Middleton, R.H., Fu, M.: Further results on localization based switching adaptive control. Automatica 37, 257–263 (2001)
[120] Zhou, K., Doyle, J.C., Glover, K.: Robust and Optimal Control. Prentice Hall, Englewood Cliffs (1996)

# Index

# Lecture Notes in Control and Information Sciences

**Edited by M. Thoma, F. Allgöwer, M. Morari**

Further volumes of this series can be found on our homepage:
springer.com

**Vol. 405:** Stefanovic, M., Safonov, M.G.;
Safe Adaptive Control
145 p. 2011 [978-1-84996-452-4]

**Vol. 404:** Giri, F.; Bai, E.-W. (Eds.):
Block-oriented Nonlinear System Identification
425 p. 2010 [978-1-84996-512-5]

**Vol. 403:** Tóth, R.;
Modeling and Identification of
Linear Parameter-Varying Systems
319 p. 2010 [978-3-642-13811-9]

**Vol. 402:** del Re, L.; Allgöwer, F.;
Glielmo, L.; Guardiola, C.;
Kolmanovsky, I. (Eds.):
Automotive Model Predictive Control
284 p. 2010 [978-1-84996-070-0]

**Vol. 401:** Chesi, G.; Hashimoto, K. (Eds.):
Visual Servoing via Advanced
Numerical Methods
393 p. 2010 [978-1-84996-088-5]

**Vol. 400:** Tomás-Rodríguez, M.;
Banks, S.P.:
Linear, Time-varying Approximations
to Nonlinear Dynamical Systems
298 p. 2010 [978-1-84996-100-4]

**Vol. 399:** Edwards, C.; Lombaerts, T.;
Smaili, H. (Eds.):
Fault Tolerant Flight Control
appro. 350 p. 2010 [978-3-642-11689-6]

**Vol. 398:** Hara, S.; Ohta, Y.;
Willems, J.C.; Hisaya, F. (Eds.):
Perspectives in Mathematical System
Theory, Control, and Signal Processing
appro. 370 p. 2010 [978-3-540-93917-7]

**Vol. 397:** Yang, H.; Jiang, B.;
Cocquempot, V.:
Fault Tolerant Control Design for
Hybrid Systems
191 p. 2010 [978-3-642-10680-4]

**Vol. 396:** Kozlowski, K. (Ed.):
Robot Motion and Control 2009
475 p. 2009 [978-1-84882-984-8]

**Vol. 395:** Talebi, H.A.; Abdollahi, F.;
Patel, R.V.; Khorasani, K.:
Neural Network-Based State
Estimation of Nonlinear Systems
appro. 175 p. 2010 [978-1-4419-1437-8]

**Vol. 394:** Pipeleers, G.; Demeulenaere, B.;
Swevers, J.:
Optimal Linear Controller Design for
Periodic Inputs
177 p. 2009 [978-1-84882-974-9]

**Vol. 393:** Ghosh, B.K.; Martin, C.F.;
Zhou, Y.:
Emergent Problems in Nonlinear
Systems and Control
285 p. 2009 [978-3-642-03626-2]

**Vol. 392:** Bandyopadhyay, B.;
Deepak, F.; Kim, K.-S.:
Sliding Mode Control Using Novel Sliding
Surfaces
137 p. 2009 [978-3-642-03447-3]

**Vol. 391:** Khaki-Sedigh, A.; Moaveni, B.:
Control Configuration Selection for
Multivariable Plants
232 p. 2009 [978-3-642-03192-2]

**Vol. 390:** Chesi, G.; Garulli, A.;
Tesi, A.; Vicino, A.:
Homogeneous Polynomial Forms for
Robustness Analysis of Uncertain
Systems
197 p. 2009 [978-1-84882-780-6]

**Vol. 389:** Bru, R.; Romero-Vivó,
S. (Eds.):
Positive Systems
398 p. 2009 [978-3-642-02893-9]

**Vol. 388:** Jacques Loiseau, J.; Michiels, W.;
Niculescu, S-I.; Sipahi, R. (Eds.):
Topics in Time Delay Systems
418 p. 2009 [978-3-642-02896-0]

**Vol. 387:** Xia, Y.;
Fu, M.; Shi, P.:
Analysis and Synthesis of
Dynamical Systems with Time-Delays
283 p. 2009 [978-3-642-02695-9]

**Vol. 386:** Huang, D.;
Nguang, S.K.:
Robust Control for Uncertain
Networked Control Systems with
Random Delays
159 p. 2009 [978-1-84882-677-9]

**Vol. 385:** Jungers, R.:
The Joint Spectral Radius
144 p. 2009 [978-3-540-95979-3]

**Vol. 384:** Magni, L.; Raimondo, D.M.;
Allgöwer, F. (Eds.):
Nonlinear Model Predictive Control
572 p. 2009 [978-3-642-01093-4]

**Vol. 383:** Sobhani-Tehrani E.;
Khorasani K.;
Fault Diagnosis of Nonlinear Systems
Using a Hybrid Approach
XXX p. 2009 [978-0-387-92906-4]

**Vol. 382:** Bartoszewicz A.;
Nowacka-Leverton A.;
Time-Varying Sliding Modes for Second
and Third Order Systems
192 p. 2009 [978-3-540-92216-2]

**Vol. 381:** Hirsch M.J.; Commander C.W.;
Pardalos P.M.; Murphey R. (Eds.)
Optimization and Cooperative Control Strategies:
Proceedings of the 8th International Conference
on Cooperative Control and Optimization
459 p. 2009 [978-3-540-88062-2]

**Vol. 380:** Basin M.
New Trends in Optimal Filtering and Control for
Polynomial and Time-Delay Systems
206 p. 2008 [978-3-540-70802-5]

**Vol. 379:** Mellodge P.; Kachroo P.;
Model Abstraction in Dynamical Systems:
Application to Mobile Robot Control
116 p. 2008 [978-3-540-70792-9]

**Vol. 378:** Femat R.; Solis-Perales G.;
Robust Synchronization of Chaotic Systems
Via Feedback
199 p. 2008 [978-3-540-69306-2]

**Vol. 377:** Patan K.
Artificial Neural Networks for
the Modelling and Fault
Diagnosis of Technical Processes
206 p. 2008 [978-3-540-79871-2]

**Vol. 376:** Hasegawa Y.
Approximate and Noisy Realization of
Discrete-Time Dynamical Systems
245 p. 2008 [978-3-540-79433-2]

**Vol. 375:** Bartolini G.; Fridman L.; Pisano A.;
Usai E. (Eds.)
Modern Sliding Mode Control Theory
465 p. 2008 [978-3-540-79015-0]

**Vol. 374:** Huang B.; Kadali R.
Dynamic Modeling, Predictive Control
and Performance Monitoring
240 p. 2008 [978-1-84800-232-6]

**Vol. 373:** Wang Q.-G.; Ye Z.; Cai W.-J.;
Hang C.-C.
PID Control for Multivariable Processes
264 p. 2008 [978-3-540-78481-4]

**Vol. 372:** Zhou J.; Wen C.
Adaptive Backstepping Control of Uncertain
Systems
241 p. 2008 [978-3-540-77806-6]

**Vol. 371:** Blondel V.D.; Boyd S.P.;
Kimura H. (Eds.)
Recent Advances in Learning and Control
279 p. 2008 [978-1-84800-154-1]

**Vol. 370:** Lee S.; Suh I.H.;
Kim M.S. (Eds.)
Recent Progress in Robotics:
Viable Robotic Service to Human
410 p. 2008 [978-3-540-76728-2]

**Vol. 369:** Hirsch M.J.; Pardalos P.M.;
Murphey R.; Grundel D.
Advances in Cooperative Control and
Optimization
423 p. 2007 [978-3-540-74354-5]

**Vol. 368:** Chee F.; Fernando T.
Closed-Loop Control of Blood Glucose
157 p. 2007 [978-3-540-74030-8]

**Vol. 367:** Turner M.C.; Bates D.G. (Eds.)
Mathematical Methods for Robust and Nonlinear
Control
444 p. 2007 [978-1-84800-024-7]

**Vol. 366:** Bullo F.; Fujimoto K. (Eds.)
Lagrangian and Hamiltonian Methods for
Nonlinear Control 2006
398 p. 2007 [978-3-540-73889-3]

**Vol. 365:** Bates D.; Hagström M. (Eds.)
Nonlinear Analysis and Synthesis Techniques
for Aircraft Control
360 p. 2007 [978-3-540-73718-6]

**Vol. 364:** Chiuso A.; Ferrante A.;
Pinzoni S. (Eds.)
Modeling, Estimation and Control
356 p. 2007 [978-3-540-73569-4]

**Vol. 363:** Besançon G. (Ed.)
Nonlinear Observers and Applications
224 p. 2007 [978-3-540-73502-1]